# Hadoop MapReduce v2 Cookbook
## Second Edition

Explore the Hadoop MapReduce v2 ecosystem to gain insights from very large datasets

**Thilina Gunarathne**

[PACKT] open source*
PUBLISHING   community experience distilled

BIRMINGHAM - MUMBAI

# Hadoop MapReduce v2 Cookbook
## *Second Edition*

First published: January 2013

Second edition: February 2015

Production reference: 1200215

Published by Packt Publishing Ltd.
Livery Place
35 Livery Street
Birmingham B3 2PB, UK.

ISBN 978-1-78328-547-1

www.packtpub.com

Cover image by Jarek Blaminsky (milak6@wp.pl)

# Credits

**Authors**
Thilina Gunarathne
Srinath Perera

**Reviewers**
Skanda Bhargav
Randal Scott King
Dmitry Spikhalskiy
Jeroen van Wilgenburg
Shinichi Yamashita

**Commissioning Editor**
Edward Gordon

**Acquisition Editors**
Joanne Fitzpatrick

**Content Development Editor**
Shweta Pant

**Technical Editors**
Indrajit A. Das
Pankaj Kadam

**Copy Editors**
Puja Lalwani
Alfida Paiva
Laxmi Subramanian

**Project Coordinator**
Shipra Chawhan

**Proofreaders**
Bridget Braund
Maria Gould
Paul Hindle
Bernadette Watkins

**Indexer**
Priya Sane

**Production Coordinator**
Nitesh Thakur

**Cover Work**
Nitesh Thakur

# About the Author

**Thilina Gunarathne** is a senior data scientist at KPMG LLP. He led the Hadoop-related efforts at Link Analytics before its acquisition by KPMG LLP. He has extensive experience in using Apache Hadoop and its related technologies for large-scale data-intensive computations. He coauthored the first edition of this book, *Hadoop MapReduce Cookbook*, with Dr. Srinath Perera.

Thilina has contributed to several open source projects at Apache Software Foundation as a member, committer, and a PMC member. He has also published many peer-reviewed research articles on how to extend the MapReduce model to perform efficient data mining and data analytics computations in the cloud. Thilina received his PhD and MSc degrees in computer science from Indiana University, Bloomington, USA, and received his bachelor of science degree in computer science and engineering from University of Moratuwa, Sri Lanka.

# Acknowledgments

I would like to thank my wife, Bimalee, my son, Kaveen, and my daughter, Yasali, for putting up with me for all the missing family time and for providing me with love and encouragement throughout the writing period. I would also like to thank my parents and siblings. Without their love, guidance, and encouragement, I would not be where I am today.

I really appreciate the contributions from my coauthor, Dr. Srinath Perera, for the first edition of this book. Many of his contributions from the first edition of this book have been adapted to the current book even though he wasn't able to coauthor this book due to his work and family commitments.

I would like to thank the Hadoop, HBase, Mahout, Pig, Hive, Sqoop, Nutch, and Lucene communities for developing great open source products. Thanks to Apache Software Foundation for fostering vibrant open source communities.

Big thanks to the editorial staff at Packt for providing me with the opportunity to write this book and feedback and guidance throughout the process. Thanks to the reviewers of this book for the many useful suggestions and corrections.

I would like to express my deepest gratitude to all the mentors I have had over the years, including Prof. Geoffrey Fox, Dr. Chris Groer, Dr. Sanjiva Weerawarana, Prof. Dennis Gannon, Prof. Judy Qiu, Prof. Beth Plale, and all my professors at Indiana University and University of Moratuwa for all the knowledge and guidance they gave me. Thanks to all my past and present colleagues for the many insightful discussions we've had and the knowledge they shared with me.

# About the Author

**Srinath Perera** (coauthor of the first edition of this book) is a senior software architect at WSO2 Inc., where he overlooks the overall WSO2 platform architecture with the CTO. He also serves as a research scientist at Lanka Software Foundation and teaches as a member of the visiting faculty at Department of Computer Science and Engineering, University of Moratuwa. He is a cofounder of Apache Axis2 open source project, and he has been involved with the Apache Web Service project since 2002 and is a member of Apache Software foundation and Apache Web Service project PMC. Srinath is also a committer of Apache open source projects Axis, Axis2, and Geronimo.

Srinath received his PhD and MSc in computer science from Indiana University, Bloomington, USA, and his bachelor of science in computer science and engineering from University of Moratuwa, Sri Lanka.

Srinath has authored many technical and peer-reviewed research articles; more details can be found on his website. He is also a frequent speaker at technical venues.

Srinath has worked with large-scale distributed systems for a long time. He closely works with big data technologies such as Hadoop and Cassandra daily. He also teaches a parallel programming graduate class at University of Moratuwa, which is primarily based on Hadoop.

I would like to thank my wife, Miyuru, and my parents, whose never-ending support keeps me going. I would also like to thank Sanjiva from WSO2 who encouraged us to make our mark even though project such as these are not in the job description. Finally, I would like to thank my colleagues at WSO2 for ideas and companionship that have shaped the book in many ways.

# About the Reviewers

**Skanda Bhargav** is an engineering graduate from Visvesvaraya Technological University (VTU), Belgaum, Karnataka, India. He did his majors in computer science engineering. He is currently employed with Happiest Minds Technologies, an MNC based out of Bangalore. He is a Cloudera-certified developer in Apache Hadoop. His interests are big data and Hadoop.

He has been a reviewer for the following books and a video, all by Packt Publishing:

- *Instant MapReduce Patterns – Hadoop Essentials How-to*
- *Hadoop Cluster Deployment*
- *Building Hadoop Clusters [Video]*
- *Cloudera Administration Handbook*

I would like to thank my family for their immense support and faith in me throughout my learning stage. My friends have brought the confidence in me to a level that makes me bring out the best in myself. I am happy that God has blessed me with such wonderful people, without whom I wouldn't have tasted the success that I've achieved today.

**Randal Scott King** is a global consultant who specializes in big data and network architecture. His 15 years of experience in IT consulting has resulted in a client list that looks like a "Who's Who" of the Fortune 500. His recent projects include a complete network redesign for an aircraft manufacturer and an in-store video analytics pilot for a major home improvement retailer.

He lives with his children outside Atlanta, GA. You can visit his blog at www.randalscottking.com.

**Dmitry Spikhalskiy** currently holds the position of software engineer in a Russian social network service, Odnoklassniki, and is working on a search engine, video recommendation system, and movie content analysis.

Previously, Dmitry took part in developing Mind Labs' platform, infrastructure, and benchmarks for a high-load video conference and streaming service, which got "The biggest online-training in the world" Guinness world record with more than 12,000 participants. As a technical lead and architect, he launched a mobile social banking start-up called Instabank. He has also reviewed *Learning Google Guice* and *PostgreSQL 9 Admin Cookbook*, both by Packt Publishing.

Dmitry graduated from Moscow State University with an MSc degree in computer science, where he first got interested in parallel data processing, high-load systems, and databases.

**Jeroen van Wilgenburg** is a software craftsman at JPoint (`http://www.jpoint.nl`), a software agency based in the center of the Netherlands. Their main focus is on developing high-quality Java and Scala software with open source frameworks.

Currently, Jeroen is developing several big data applications with Hadoop, MapReduce, Storm, Spark, Kafka, MongoDB, and Elasticsearch.

Jeroen is a car enthusiast and likes to be outdoors, usually training for a triathlon. In his spare time, Jeroen writes about his work experience at `http://vanwilgenburg.wordpress.com`.

**Shinichi Yamashita** is a solutions architect at System Platform Sector in NTT DATA Corporation, Japan. He has more than 9 years of experience in software and middleware engineering (Apache, Tomcat, PostgreSQL, Hadoop Ecosystem, and Spark). Shinichi has written a few books on Hadoop in Japanese.

# www.PacktPub.com

## Support files, eBooks, discount offers, and more

For support files and downloads related to your book, please visit www.PacktPub.com.

Did you know that Packt offers eBook versions of every book published, with PDF and ePub files available? You can upgrade to the eBook version at www.PacktPub.com and as a print book customer, you are entitled to a discount on the eBook copy. Get in touch with us at service@packtpub.com for more details.

At www.PacktPub.com, you can also read a collection of free technical articles, sign up for a range of free newsletters and receive exclusive discounts and offers on Packt books and eBooks.

https://www2.packtpub.com/books/subscription/packtlib

Do you need instant solutions to your IT questions? PacktLib is Packt's online digital book library. Here, you can search, access, and read Packt's entire library of books.

## Why Subscribe?

- ► Fully searchable across every book published by Packt
- ► Copy and paste, print, and bookmark content
- ► On demand and accessible via a web browser

## Free Access for Packt account holders

If you have an account with Packt at www.PacktPub.com, you can use this to access PacktLib today and view 9 entirely free books. Simply use your login credentials for immediate access.

# Table of Contents

# Preface

We are currently facing an avalanche of data, and this data contains many insights that hold the keys to success or failure in the data-driven world. Next generation Hadoop (v2) offers a cutting-edge platform to store and analyze these massive data sets and improve upon the widely used and highly successful Hadoop MapReduce v1. The recipes that will help you analyze large and complex datasets with next generation Hadoop MapReduce will provide you with the skills and knowledge needed to process large and complex datasets using the next generation Hadoop ecosystem.

This book presents many exciting topics such as MapReduce patterns using Hadoop to solve analytics, classifications, and data indexing and searching. You will also be introduced to several Hadoop ecosystem components including Hive, Pig, HBase, Mahout, Nutch, and Sqoop.

This book introduces you to simple examples and then dives deep to solve in-depth big data use cases. This book presents more than 90 ready-to-use Hadoop MapReduce recipes in a simple and straightforward manner, with step-by-step instructions and real-world examples.

## What this book covers

*Chapter 1, Getting Started with Hadoop v2*, introduces Hadoop MapReduce, YARN, and HDFS, and walks through the installation of Hadoop v2.

*Chapter 2, Cloud Deployments – Using Hadoop Yarn on Cloud Environments*, explains how to use Amazon Elastic MapReduce (EMR) and Apache Whirr to deploy and execute Hadoop MapReduce, Pig, Hive, and HBase computations on cloud infrastructures.

*Chapter 3, Hadoop Essentials – Configurations, Unit Tests, and Other APIs*, introduces basic Hadoop YARN and HDFS configurations, HDFS Java API, and unit testing methods for MapReduce applications.

*Chapter 4, Developing Complex Hadoop MapReduce Applications*, introduces you to several advanced Hadoop MapReduce features that will help you develop highly customized and efficient MapReduce applications.

*Chapter 5, Analytics*, explains how to perform basic data analytic operations using Hadoop MapReduce.

*Chapter 6, Hadoop Ecosystem – Apache Hive*, introduces Apache Hive, which provides data warehouse capabilities on top of Hadoop, using a SQL-like query language.

*Chapter 7, Hadoop Ecosystem II – Pig, HBase, Mahout, and Sqoop*, introduces the Apache Pig data flow style data-processing language, Apache HBase NoSQL data storage, Apache Mahout machine learning and data-mining toolkit, and Apache Sqoop bulk data transfer utility to transfer data between Hadoop and the relational databases.

*Chapter 8, Searching and Indexing*, introduces several tools and techniques that you can use with Apache Hadoop to perform large-scale searching and indexing.

*Chapter 9, Classifications, Recommendations, and Finding Relationships*, explains how to implement complex algorithms such as classifications, recommendations, and finding relationships using Hadoop.

*Chapter 10, Mass Text Data Processing*, explains how to use Hadoop and Mahout to process large text datasets and how to perform data preprocessing and loading of operations using Hadoop.

# What you need for this book

You need a moderate knowledge of Java and access to the Internet and a computer that runs a Linux operating system.

# Who this book is for

If you are a big data enthusiast and wish to use Hadoop v2 to solve your problems, then this book is for you. This book is for Java programmers with little to moderate knowledge of Hadoop MapReduce. This is also a one-stop reference for developers and system admins who want to quickly get up to speed with using Hadoop v2. It would be helpful to have a basic knowledge of software development using Java and a basic working knowledge of Linux.

# Conventions

In this book, you will find a number of styles of text that distinguish between different kinds of information. Here are some examples of these styles, and an explanation of their meaning.

Code words in text, database table names, folder names, filenames, file extensions, pathnames, dummy URLs, user input, and Twitter handles are shown as follows: "The following are the descriptions of the properties we used in the `hadoop.properties` file."

A block of code is set as follows:

```
Path file = new Path("demo.txt");
FSDataOutputStream outStream = fs.create(file);
outStream.writeUTF("Welcome to HDFS Java API!!!");
outStream.close();
```

When we wish to draw your attention to a particular part of a code block, the relevant lines or items are set in bold:

```
Job job = Job.getInstance(getConf(), "MLReceiveReplyProcessor");
job.setJarByClass(CountReceivedRepliesMapReduce.class);
job.setMapperClass(AMapper.class);
job.setReducerClass(AReducer.class);
job.setNumReduceTasks(numReduce);

job.setOutputKeyClass(Text.class);
job.setOutputValueClass(Text.class);
job.setInputFormatClass(MBoxFileInputFormat.class);
FileInputFormat.setInputPaths(job, new Path(inputPath));
FileOutputFormat.setOutputPath(job, new Path(outputPath));

int exitStatus = job.waitForCompletion(true) ? 0 : 1;
```

Any command-line input or output is written as follows:

```
205.212.115.106 - - [01/Jul/1995:00:00:12 -0400] "GET
/shuttle/countdown/countdown.html HTTP/1.0" 200 3985
```

**New terms** and **important words** are shown in bold. Words that you see on the screen, in menus or dialog boxes for example, appear in the text like this: "Select **Custom Action** in the **Add Bootstrap Actions** drop-down box. Click on **Configure and add**."

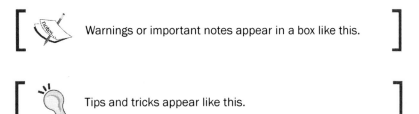

Warnings or important notes appear in a box like this.

Tips and tricks appear like this.

# Reader feedback

Feedback from our readers is always welcome. Let us know what you think about this book—what you liked or may have disliked. Reader feedback is important for us to develop titles that you really get the most out of.

To send us general feedback, simply send an e-mail to feedback@packtpub.com, and mention the book title via the subject of your message.

If there is a topic that you have expertise in and you are interested in either writing or contributing to a book, see our author guide on www.packtpub.com/authors.

# Customer support

Now that you are the proud owner of a Packt book, we have a number of things to help you to get the most from your purchase.

## Downloading the example code

You can download the example code files for all Packt books you have purchased from your account at http://www.packtpub.com. If you purchased this book elsewhere, you can visit http://www.packtpub.com/support and register to have the files e-mailed directly to you.

## Errata

Although we have taken every care to ensure the accuracy of our content, mistakes do happen. If you find a mistake in one of our books—maybe a mistake in the text or the code—we would be grateful if you would report this to us. By doing so, you can save other readers from frustration and help us improve subsequent versions of this book. If you find any errata, please report them by visiting http://www.packtpub.com/submit-errata, selecting your book, clicking on the **errata submission form** link, and entering the details of your errata. Once your errata are verified, your submission will be accepted and the errata will be uploaded on our website, or added to any list of existing errata, under the Errata section of that title. Any existing errata can be viewed by selecting your title from http://www.packtpub.com/support.

## Piracy

Piracy of copyright material on the Internet is an ongoing problem across all media. At Packt, we take the protection of our copyright and licenses very seriously. If you come across any illegal copies of our works, in any form, on the Internet, please provide us with the location address or website name immediately so that we can pursue a remedy.

Please contact us at copyright@packtpub.com with a link to the suspected pirated material.

We appreciate your help in protecting our authors, and our ability to bring you valuable content.

## Questions

You can contact us at questions@packtpub.com if you are having a problem with any aspect of the book, and we will do our best to address it.

# 1
# Getting Started with Hadoop v2

In this chapter, we will cover the following recipes:

- ▸ Setting up standalone Hadoop v2 on your local machine
- ▸ Writing a WordCount MapReduce application, bundling it, and running it using Hadoop local mode
- ▸ Adding a combiner step to the WordCount MapReduce program
- ▸ Setting up HDFS
- ▸ Setting up Hadoop YARN in a distributed cluster environment using Hadoop v2
- ▸ Setting up Hadoop ecosystem in a distributed cluster environment using a Hadoop distribution
- ▸ HDFS command-line file operations
- ▸ Running the WordCount program in a distributed cluster environment
- ▸ Benchmarking HDFS using DFSIO
- ▸ Benchmarking Hadoop MapReduce using TeraSort

## Introduction

We are living in the era of big data, where exponential growth of phenomena such as web, social networking, smartphones, and so on are producing petabytes of data on a daily basis. Gaining insights from analyzing these very large amounts of data has become a *must-have* competitive advantage for many industries. However, the size and the possibly unstructured nature of these data sources make it impossible to use traditional solutions such as relational databases to store and analyze these datasets.

Storage, processing, and analyzing petabytes of data in a meaningful and timely manner require many compute nodes with thousands of disks and thousands of processors together with the ability to efficiently communicate massive amounts of data among them. Such a scale makes failures such as disk failures, compute node failures, network failures, and so on a common occurrence making fault tolerance a very important aspect of such systems. Other common challenges that arise include the significant cost of resources, handling communication latencies, handling heterogeneous compute resources, synchronization across nodes, and load balancing. As you can infer, developing and maintaining distributed parallel applications to process massive amounts of data while handling all these issues is not an easy task. This is where Apache Hadoop comes to our rescue.

Google is one of the first organizations to face the problem of processing massive amounts of data. Google built a framework for large-scale data processing borrowing the **map** and **reduce** paradigms from the functional programming world and named it as **MapReduce**. At the foundation of Google, MapReduce was the Google File System, which is a high throughput parallel filesystem that enables the reliable storage of massive amounts of data using commodity computers. Seminal research publications that introduced Google MapReduce and Google File System concepts can be found at `http://research.google.com/archive/mapreduce.html` and `http://research.google.com/archive/gfs.html`.

Apache Hadoop MapReduce is the most widely known and widely used open source implementation of the Google MapReduce paradigm. Apache **Hadoop Distributed File System** (**HDFS**) provides an open source implementation of the Google File Systems concept.

Apache Hadoop MapReduce, HDFS, and YARN provide a scalable, fault-tolerant, distributed platform for storage and processing of very large datasets across clusters of commodity computers. Unlike in traditional **High Performance Computing** (**HPC**) clusters, Hadoop uses the same set of compute nodes for data storage as well as to perform the computations, allowing Hadoop to improve the performance of large scale computations by collocating computations with the storage. Also, the hardware cost of a Hadoop cluster is orders of magnitude cheaper than HPC clusters and database appliances due to the usage of commodity hardware and commodity interconnects. Together Hadoop-based frameworks have become the de-facto standard for storing and processing big data.

# Hadoop Distributed File System – HDFS

HDFS is a block structured distributed filesystem that is designed to store petabytes of data reliably on compute clusters made out of commodity hardware. HDFS overlays on top of the existing filesystem of the compute nodes and stores files by breaking them into coarser grained blocks (for example, 128 MB). HDFS performs better with large files. HDFS distributes the data blocks of large files across to all the nodes of the cluster to facilitate the very high parallel aggregate read bandwidth when processing the data. HDFS also stores redundant copies of these data blocks in multiple nodes to ensure reliability and fault tolerance. Data processing frameworks such as MapReduce exploit these distributed sets of data blocks and the redundancy to maximize the data local processing of large datasets, where most of the data blocks would get processed locally in the same physical node as they are stored.

HDFS consists of **NameNode** and **DataNode** services providing the basis for the distributed filesystem. NameNode stores, manages, and serves the metadata of the filesystem. NameNode does not store any real data blocks. DataNode is a per node service that manages the actual data block storage in the DataNodes. When retrieving data, client applications first contact the NameNode to get the list of locations the requested data resides in and then contact the DataNodes directly to retrieve the actual data. The following diagram depicts a high-level overview of the structure of HDFS:

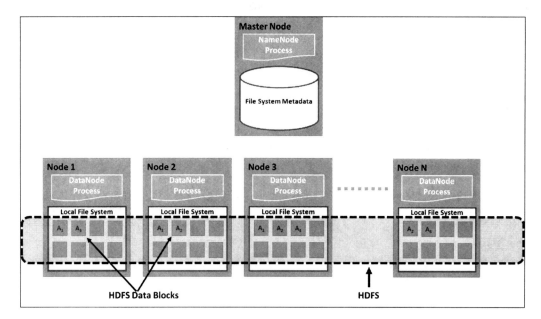

Hadoop v2 brings in several performance, scalability, and reliability improvements to HDFS. One of the most important among those is the **High Availability** (**HA**) support for the HDFS NameNode, which provides manual and automatic failover capabilities for the HDFS NameNode service. This solves the widely known NameNode single point of failure weakness of HDFS. Automatic NameNode high availability of Hadoop v2 uses Apache ZooKeeper for failure detection and for active NameNode election. Another important new feature is the support for HDFS federation. HDFS federation enables the usage of multiple independent HDFS namespaces in a single HDFS cluster. These namespaces would be managed by independent NameNodes, but share the DataNodes of the cluster to store the data. The HDFS federation feature improves the horizontal scalability of HDFS by allowing us to distribute the workload of NameNodes. Other important improvements of HDFS in Hadoop v2 include the support for HDFS snapshots, heterogeneous storage hierarchy support (Hadoop 2.3 or higher), in-memory data caching support (Hadoop 2.3 or higher), and many performance improvements.

Almost all the Hadoop ecosystem data processing technologies utilize HDFS as the primary data storage. HDFS can be considered as the most important component of the Hadoop ecosystem due to its central nature in the Hadoop architecture.

## Hadoop YARN

**YARN** (**Yet Another Resource Negotiator**) is the major new improvement introduced in Hadoop v2. YARN is a resource management system that allows multiple distributed processing frameworks to effectively share the compute resources of a Hadoop cluster and to utilize the data stored in HDFS. YARN is a central component in the Hadoop v2 ecosystem and provides a common platform for many different types of distributed applications.

The batch processing based MapReduce framework was the only natively supported data processing framework in Hadoop v1. While MapReduce works well for analyzing large amounts of data, MapReduce by itself is not sufficient enough to support the growing number of other distributed processing use cases such as real-time data computations, graph computations, iterative computations, and real-time data queries. The goal of YARN is to allow users to utilize multiple distributed application frameworks that provide such capabilities side by side sharing a single cluster and the HDFS filesystem. Some examples of the current YARN applications include the MapReduce framework, Tez high performance processing framework, Spark processing engine, and the Storm real-time stream processing framework. The following diagram depicts the high-level architecture of the YARN ecosystem:

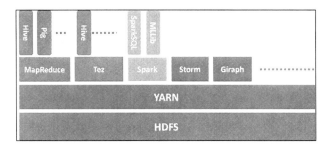

The YARN ResourceManager process is the central resource scheduler that manages and allocates resources to the different applications (also known as jobs) submitted to the cluster. YARN NodeManager is a per node process that manages the resources of a single compute node. Scheduler component of the ResourceManager allocates resources in response to the resource requests made by the applications, taking into consideration the cluster capacity and the other scheduling policies that can be specified through the YARN policy plugin framework.

YARN has a concept called containers, which is the unit of resource allocation. Each allocated container has the rights to a certain amount of CPU and memory in a particular compute node. Applications can request resources from YARN by specifying the required number of containers and the CPU and memory required by each container.

ApplicationMaster is a per-application process that coordinates the computations for a single application. The first step of executing a YARN application is to deploy the ApplicationMaster. After an application is submitted by a YARN client, the ResourceManager allocates a container and deploys the ApplicationMaster for that application. Once deployed, the ApplicationMaster is responsible for requesting and negotiating the necessary resource containers from the ResourceManager. Once the resources are allocated by the ResourceManager, ApplicationMaster coordinates with the NodeManagers to launch and monitor the application containers in the allocated resources. The shifting of application coordination responsibilities to the ApplicationMaster reduces the burden on the ResourceManager and allows it to focus solely on managing the cluster resources. Also having separate ApplicationMasters for each submitted application improves the scalability of the cluster as opposed to having a single process bottleneck to coordinate all the application instances. The following diagram depicts the interactions between various YARN components, when a MapReduce application is submitted to the cluster:

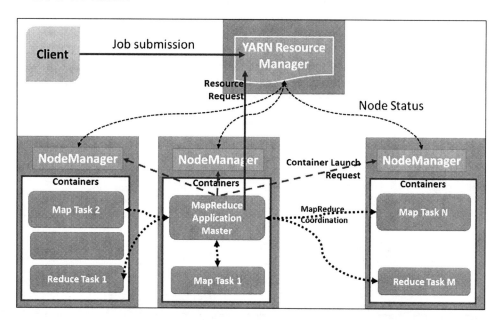

While YARN supports many different distributed application execution frameworks, our focus in this book is mostly on traditional MapReduce and related technologies.

# Hadoop MapReduce

Hadoop MapReduce is a data processing framework that can be utilized to process massive amounts of data stored in HDFS. As we mentioned earlier, distributed processing of a massive amount of data in a reliable and efficient manner is not an easy task. Hadoop MapReduce aims to make it easy for users by providing a clean abstraction for programmers by providing automatic parallelization of the programs and by providing framework managed fault tolerance support.

MapReduce programming model consists of Map and Reduce functions. The Map function receives each record of the input data (lines of a file, rows of a database, and so on) as key-value pairs and outputs key-value pairs as the result. By design, each Map function invocation is independent of each other allowing the framework to use divide and conquer to execute the computation in parallel. This also allows duplicate executions or re-executions of the Map tasks in case of failures or load imbalances without affecting the results of the computation. Typically, Hadoop creates a single Map task instance for each HDFS data block of the input data. The number of Map function invocations inside a Map task instance is equal to the number of data records in the input data block of the particular Map task instance.

Hadoop MapReduce groups the output key-value records of all the Map tasks of a computation by the **key** and distributes them to the Reduce tasks. This distribution and transmission of data to the Reduce tasks is called the Shuffle phase of the MapReduce computation. Input data to each Reduce task would also be sorted and grouped by the key. The Reduce function gets invoked for each key and the group of values of that key (*reduce <key, list_of_values>*) in the sorted order of the keys. In a typical MapReduce program, users only have to implement the Map and Reduce functions and Hadoop takes care of scheduling and executing them in parallel. Hadoop will rerun any failed tasks and also provide measures to mitigate any unbalanced computations. Have a look at the following diagram for a better understanding of the MapReduce data and computational flows:

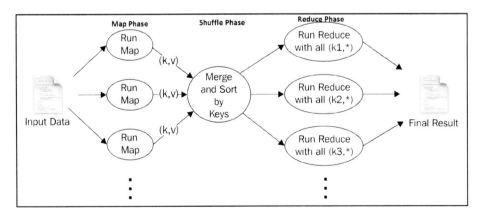

In Hadoop 1.x, the MapReduce (MR1) components consisted of the **JobTracker** process, which ran on a master node managing the cluster and coordinating the jobs, and **TaskTrackers**, which ran on each compute node launching and coordinating the tasks executing in that node. Neither of these processes exist in Hadoop 2.x MapReduce (MR2). In MR2, the job coordinating responsibility of JobTracker is handled by an ApplicationMaster that will get deployed on-demand through YARN. The cluster management and job scheduling responsibilities of JobTracker are handled in MR2 by the YARN ResourceManager. JobHistoryServer has taken over the responsibility of providing information about the completed MR2 jobs. YARN NodeManagers provide the functionality that is somewhat similar to MR1 TaskTrackers by managing resources and launching containers (which in the case of MapReduce 2 houses Map or Reduce tasks) in the compute nodes.

## Hadoop installation modes

Hadoop v2 provides three installation choices:

- ▶ **Local mode**: The local mode allows us to run MapReduce computation using just the unzipped Hadoop distribution. This nondistributed mode executes all parts of Hadoop MapReduce within a single Java process and uses the local filesystem as the storage. The local mode is very useful for testing/debugging the MapReduce applications locally.

- ▶ **Pseudo distributed mode**: Using this mode, we can run Hadoop on a single machine emulating a distributed cluster. This mode runs the different services of Hadoop as different Java processes, but within a single machine. This mode is good to let you play and experiment with Hadoop.

- ▶ **Distributed mode**: This is the real distributed mode that supports clusters that span from a few nodes to thousands of nodes. For production clusters, we recommend using one of the many packaged Hadoop distributions as opposed to installing Hadoop from scratch using the Hadoop release binaries, unless you have a specific use case that requires a vanilla Hadoop installation. Refer to the *Setting up Hadoop ecosystem in a distributed cluster environment using a Hadoop distribution* recipe for more information on Hadoop distributions.

The example code files for this book are available on GitHub at `https://github.com/thilg/hcb-v2`. The `chapter1` folder of the code repository contains the sample source code files for this chapter. You can also download all the files in the repository using the `https://github.com/thilg/hcb-v2/archive/master.zip` link.

The sample code for this book uses Gradle to automate the compiling and building of the projects. You can install Gradle by following the guide provided at `http://www.gradle.org/docs/current/userguide/installation.html`. Usually, you only have to download and extract the Gradle distribution from `http://www.gradle.org/downloads` and add the bin directory of the extracted Gradle distribution to your path variable.

All the sample code can be built by issuing the `gradle build` command in the main folder of the code repository.

Project files for Eclipse IDE can be generated by running the `gradle eclipse` command in the main folder of the code repository.

Project files for the IntelliJ IDEA IDE can be generated by running the `gradle idea` command in the main folder of the code repository.

# Setting up Hadoop v2 on your local machine

This recipe describes how to set up Hadoop v2 on your local machine using the local mode. Local mode is a non-distributed mode that can be used for testing and debugging your Hadoop applications. When running a Hadoop application in local mode, all the required Hadoop components and your applications execute inside a single **Java Virtual Machine** (**JVM**) process.

## Getting ready

Download and install JDK 1.6 or a higher version, preferably the Oracle JDK 1.7. Oracle JDK can be downloaded from `http://www.oracle.com/technetwork/java/javase/downloads/index.html`.

## How to do it...

Now let's start the Hadoop v2 installation:

1. Download the most recent Hadoop v2 branch distribution (Hadoop 2.2.0 or later) from `http://hadoop.apache.org/releases.html`.

2. Unzip the Hadoop distribution using the following command. You will have to change the `x.x.` in the filename to the actual release you have downloaded. From this point onward, we will call the unpacked Hadoop directory {HADOOP_HOME}:

```
$ tar -zxvf hadoop-2.x.x.tar.gz
```

3. Now, you can run Hadoop jobs through the {HADOOP_HOME}/bin/hadoop command, and we will elaborate on that further in the next recipe.

## How it works...

Hadoop local mode does not start any servers but does all the work within a single JVM. When you submit a job to Hadoop in local mode, Hadoop starts a JVM to execute the job. The output and the behavior of the job is the same as a distributed Hadoop job, except for the fact that the job only uses the current node to run the tasks and the local filesystem is used for the data storage. In the next recipe, we will discover how to run a MapReduce program using the Hadoop local mode.

# Writing a WordCount MapReduce application, bundling it, and running it using the Hadoop local mode

This recipe explains how to implement a simple MapReduce program to count the number of occurrences of words in a dataset. WordCount is famous as the HelloWorld equivalent for Hadoop MapReduce.

To run a MapReduce job, users should supply a `map` function, a `reduce` function, input data, and a location to store the output data. When executed, Hadoop carries out the following steps:

1. Hadoop uses the supplied **InputFormat** to break the input data into key-value pairs and invokes the `map` function for each key-value pair, providing the key-value pair as the input. When executed, the `map` function can output zero or more key-value pairs.

2. Hadoop transmits the key-value pairs emitted from the Mappers to the Reducers (this step is called Shuffle). Hadoop then sorts these key-value pairs by the key and groups together the values belonging to the same key.

3. For each distinct key, Hadoop invokes the reduce function once while passing that particular key and list of values for that key as the input.

4. The `reduce` function may output zero or more key-value pairs, and Hadoop writes them to the output data location as the final result.

## Getting ready

Select the source code for the first chapter from the source code repository for this book. Export the `$HADOOP_HOME` environment variable pointing to the root of the extracted Hadoop distribution.

## How to do it...

Now let's write our first Hadoop MapReduce program:

1. The WordCount sample uses MapReduce to count the number of word occurrences within a set of input documents. The sample code is available in the `chapter1/Wordcount.java` file of the source folder of this chapter. The code has three parts—Mapper, Reducer, and the main program.

2. The Mapper extends from the `org.apache.hadoop.mapreduce.Mapper` interface. Hadoop InputFormat provides each line in the input files as an input key-value pair to the `map` function. The `map` function breaks each line into substrings using whitespace characters such as the separator, and for each token (word) emits `(word,1)` as the output.

```
public void map(Object key, Text value, Context context)
throws IOException, InterruptedException {
  // Split the input text value to words
  StringTokenizer itr = new
    StringTokenizer(value.toString());

  // Iterate all the words in the input text value
  while (itr.hasMoreTokens()) {
    word.set(itr.nextToken());
    context.write(word, new IntWritable(1));
  }
}
```

3. Each `reduce` function invocation receives a key and all the values of that key as the input. The `reduce` function outputs the key and the number of occurrences of the key as the output.

```
public void reduce(Text key,
Iterable<IntWritable>values,Context context) throws
IOException, InterruptedException
{
  int sum = 0;
  // Sum all the occurrences of the word (key)
  for (IntWritableval : values) {
    sum += val.get();
  }
```

```
    result.set(sum);
    context.write(key, result);
}
```

4. The `main` driver program configures the MapReduce job and submits it to the Hadoop YARN cluster:

```
Configuration conf = new Configuration();
......
// Create a new job
Job job = Job.getInstance(conf, "word count");
// Use the WordCount.class file to point to the job jar
job.setJarByClass(WordCount.class);

job.setMapperClass(TokenizerMapper.class);
job.setReducerClass(IntSumReducer.class);
job.setOutputKeyClass(Text.class);
job.setOutputValueClass(IntWritable.class);

// Setting the input and output locations
FileInputFormat.addInputPath(job, new Path(otherArgs[0]));
FileOutputFormat.setOutputPath(job, newPath(otherArgs[1]));
// Submit the job and wait for it's completion
System.exit(job.waitForCompletion(true) ? 0 : 1);
```

5. Compile the sample using the Gradle build as mentioned in the introduction of this chapter by issuing the `gradle build` command from the `chapter1` folder of the sample code repository. Alternatively, you can also use the provided Apache Ant build file by issuing the `ant compile` command.

6. Run the WordCount sample using the following command. In this command, `chapter1.WordCount` is the name of the `main` class. `wc-input` is the input data directory and `wc-output` is the output path. The `wc-input` directory of the source repository contains a sample text file. Alternatively, you can copy any text file to the `wc-input` directory.

```
$ $HADOOP_HOME/bin/hadoop jar \
hcb-c1-samples.jar \
chapter1.WordCount wc-input wc-output
```

7. The output directory (`wc-output`) will have a file named `part-r-XXXXX`, which will have the count of each word in the document. Congratulations! You have successfully run your first MapReduce program.

```
$ cat wc-output/part*
```

## How it works...

In the preceding sample, MapReduce worked in the local mode without starting any servers and using the local filesystem as the storage system for inputs, outputs, and working data. The following diagram shows what happened in the WordCount program under the covers:

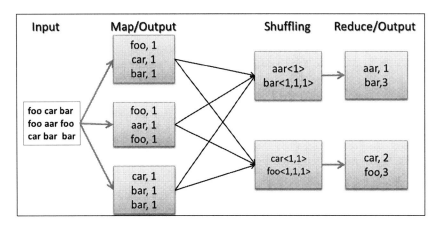

The WordCount MapReduce workflow works as follows:

1. Hadoop reads the input, breaks it using new line characters as the separator and then runs the map function passing each line as an argument with the line number as the key and the line contents as the value.

2. The map function tokenizes the line, and for each token (word), emits a key-value pair (word,1).

3. Hadoop collects all the (word,1) pairs, sorts them by the word, groups all the values emitted against each unique key, and invokes the reduce function once for each unique key passing the key and values for that key as an argument.

4. The reduce function counts the number of occurrences of each word using the values and emits it as a key-value pair.

5. Hadoop writes the final output to the output directory.

## There's more...

As an optional step, you can set up and run the WordCount application directly from your favorite **Java Integrated Development Environment (IDE)**. Project files for Eclipse IDE and IntelliJ IDEA IDE can be generated by running gradle eclipse and gradle idea commands respectively in the main folder of the code repository.

For other IDEs, you'll have to add the JAR files in the following directories to the class-path of the IDE project you create for the sample code:

- ▸ `{HADOOP_HOME}/share/hadoop/common`
- ▸ `{HADOOP_HOME}/share/hadoop/common/lib`
- ▸ `{HADOOP_HOME}/share/hadoop/mapreduce`
- ▸ `{HADOOP_HOME}/share/hadoop/yarn`
- ▸ `{HADOOP_HOME}/share/hadoop/hdfs`

Execute the `chapter1.WordCount` class by passing `wc-input` and `wc-output` as arguments. This will run the sample as before. Running MapReduce jobs from IDE in this manner is very useful for debugging your MapReduce jobs.

## See also

Although you ran the sample with Hadoop installed in your local machine, you can run it using the distributed Hadoop cluster setup with an HDFS-distributed filesystem. The *Running the WordCount program in a distributed cluster environment* recipe of this chapter will discuss how to run this sample in a distributed setup.

# Adding a combiner step to the WordCount MapReduce program

A single Map task may output many key-value pairs with the same key causing Hadoop to **shuffle** (move) all those values over the network to the Reduce tasks, incurring a significant overhead. For example, in the previous WordCount MapReduce program, when a Mapper encounters multiple occurrences of the same word in a single Map task, the `map` function would output many `<word, 1>` intermediate key-value pairs to be transmitted over the network. However, we can optimize this scenario if we can sum all the instances of `<word, 1>` pairs to a single `<word, count>` pair before sending the data across the network to the Reducers.

To optimize such scenarios, Hadoop supports a special function called **combiner**, which performs local aggregation of the Map task output key-value pairs. When provided, Hadoop calls the combiner function on the Map task outputs before persisting the data on the disk to shuffle the Reduce tasks. This can significantly reduce the amount of data shuffled from the Map tasks to the Reduce tasks. It should be noted that the combiner is an optional step of the MapReduce flow. Even when you provide a combiner implementation, Hadoop may decide to invoke it only for a subset of the Map output data or may decide to not invoke it at all.

This recipe explains how to use a combiner with the WordCount MapReduce application introduced in the previous recipe.

## How to do it...

Now let's add a combiner to the WordCount MapReduce application:

1. The combiner must have the same interface as the `reduce` function. Output key-value pair types emitted by the combiner should match the type of the Reducer input key-value pairs. For the WordCount sample, we can reuse the WordCount `reduce` function as the combiner since the input and output data types of the WordCount `reduce` function are the same.

2. Uncomment the following line in the `WordCount.java` file to enable the combiner for the WordCount application:

   ```
   job.setCombinerClass(IntSumReducer.class);
   ```

3. Recompile the code by re-running the Gradle (`gradle build`) or the Ant build (`ant compile`).

4. Run the WordCount sample using the following command. Make sure to delete the old output directory (`wc-output`) before running the job.

   ```
   $ $HADOOP_HOME/bin/hadoop jar \
   hcb-c1-samples.jar \
   chapter1.WordCount wc-input wc-output
   ```

5. The final results will be available from the `wc-output` directory.

## How it works...

When provided, Hadoop calls the combiner function on the Map task outputs before persisting the data on the disk for shuffling to the Reduce tasks. The combiner can pre-process the data generated by the Mapper before sending it to the Reducer, thus reducing the amount of data that needs to be transferred.

In the WordCount application, combiner receives *N* number of `(word,1)` pairs as input and outputs a single `(word, N)` pair. For example, if an input processed by a Map task had 1,000 occurrences of the word "the", the Mapper will generate 1,000 `(the,1)` pairs, while the combiner will generate one `(the,1000)` pair, thus reducing the amount of data that needs to be transferred to the Reduce tasks. The following diagram show the usage of the combiner in the WordCount MapReduce application:

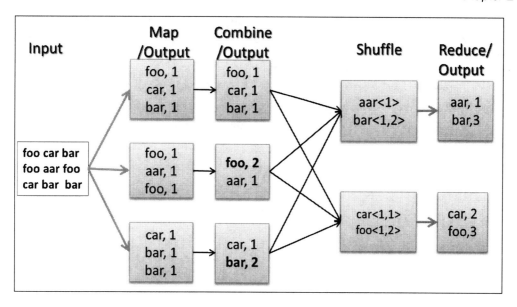

## There's more...

Using the job's `reduce` function as the combiner only works when the `reduce` function input and output key-value data types are the same. In situations where you cannot reuse the `reduce` function as the combiner, you can write a dedicated `reduce` function implementation to act as the combiner. Combiner input and output key-value pair types should be the same as the Mapper output key-value pair types.

We reiterate that the combiner is an optional step of the MapReduce flow. Even when you provide a combiner implementation, Hadoop may decide to invoke it only for a subset of the Map output data or may decide to not invoke it at all. Care should be taken not to use the combiner to perform any essential tasks of the computation as Hadoop does not guarantee the execution of the combiner.

Using a combiner does not yield significant gains in the non-distributed modes. However, in the distributed setups as described in *Setting up Hadoop YARN in a distributed cluster environment using Hadoop v2* recipe, a combiner can provide significant performance gains.

# Setting up HDFS

HDFS is a block structured distributed filesystem that is designed to store petabytes of data reliably on top of clusters made out of commodity hardware. HDFS supports storing massive amounts of data and provides high throughput access to the data. HDFS stores file data across multiple nodes with redundancy to ensure fault-tolerance and high aggregate bandwidth.

HDFS is the default distributed filesystem used by the Hadoop MapReduce computations. Hadoop supports data locality aware processing of data stored in HDFS. HDFS architecture consists mainly of a centralized NameNode that handles the filesystem metadata and DataNodes that store the real data blocks. HDFS data blocks are typically coarser grained and perform better with large streaming reads.

To set up HDFS, we first need to configure a NameNode and DataNodes, and then specify the DataNodes in the `slaves` file. When we start the NameNode, the startup script will start the DataNodes.

Installing HDFS directly using Hadoop release artifacts as mentioned in this recipe is recommended for development testing and for advanced use cases only. For regular production clusters, we recommend using a packaged Hadoop distribution as mentioned in the *Setting up Hadoop ecosystem in a distributed cluster environment using a Hadoop distribution* recipe. Packaged Hadoop distributions make it much easier to install, configure, maintain, and update the components of the Hadoop ecosystem.

## Getting ready

You can follow this recipe either using a single machine or multiple machines. If you are using multiple machines, you should choose one machine as the master node where you will run the HDFS NameNode. If you are using a single machine, use it as both the name node as well as the DataNode.

1. Install JDK 1.6 or above (Oracle JDK 1.7 is preferred) in all machines that will be used to set up the HDFS cluster. Set the `JAVA_HOME` environment variable to point to the Java installation.

2. Download Hadoop by following the *Setting up Hadoop v2 on your local machine* recipe.

## How to do it...

Now let's set up HDFS in the distributed mode:

1. Set up password-less SSH from the master node, which will be running the NameNode, to the DataNodes. Check that you can log in to localhost and to all other nodes using SSH without a passphrase by running one of the following commands:

```
$ ssh localhost
$ ssh <IPaddress>
```

**Configuring password-less SSH**

If the command in step 1 returns an error or asks for a password, create SSH keys by executing the following command (you may have to manually enable SSH beforehand depending on your OS):

```
$ ssh-keygen -t dsa -P '' -f ~/.ssh/id_dsa
```

Move the ~/.ssh/id_dsa.pub file to all the nodes in the cluster. Then add the SSH keys to the ~/.ssh/authorized_keys file in each node by running the following command (if the authorized_keys file does not exist, run the following command. Otherwise, skip to the cat command):

```
$ touch ~/.ssh/authorized_keys && chmod 600 ~/.ssh/authorized_keys
```

Now with permissions set, add your key to the ~/.ssh/authorized_keys file:

```
$ cat ~/.ssh/id_dsa.pub >> ~/.ssh/authorized_keys
```

Then you should be able to execute the following command successfully, without providing a password:

```
$ ssh localhost
```

2.  In each server, create a directory for storing HDFS data. Let's call that directory {HADOOP_DATA_DIR}. Create two subdirectories inside the data directory as {HADOOP_DATA_DIR}/data and {HADOOP_DATA_DIR}/name. Change the directory permissions to 755 by running the following command for each directory:

```
$ chmod -R 755 <HADOOP_DATA_DIR>
```

3.  In the NameNode, add the IP addresses of all the slave nodes, each on a separate line, to the {HADOOP_HOME}/etc/hadoop/slaves file. When we start the NameNode, it will use this slaves file to start the DataNodes.

4.  Add the following configurations to {HADOOP_HOME}/etc/hadoop/core-site.xml. Before adding the configurations, replace the {NAMENODE} strings with the IP of the master node:

```
<configuration>
  <property>
    <name>fs.defaultFS</name>
    <value>hdfs://{NAMENODE}:9000/</value>
  </property>
</configuration>
```

5.  Add the following configurations to the `{HADOOP_HOME}/etc/hadoop/hdfs-site.xml` files in the `{HADOOP_HOME}/etc/hadoop` directory. Before adding the configurations, replace the `{HADOOP_DATA_DIR}` with the directory you created in the first step. Replicate the `core-site.xml` and `hdfs-site.xml` files we modified in steps 4 and 5 to all the nodes.

```
<configuration>
  <property>
    <name>dfs.namenode.name.dir</name>
    <!-- Path to store namespace and transaction logs -->
    <value>{HADOOP_DATA_DIR}/name</value>
  </property>
  <property>
    <name>dfs.datanode.data.dir</name>
    <!-- Path to store data blocks in datanode -->
    <value>{HADOOP_DATA_DIR}/data</value>
  </property>
</configuration>
```

6.  From the NameNode, run the following command to format a new filesystem:

    ```
    $ $HADOOP_HOME/bin/hdfs namenode -format
    ```

    You will see the following line in the output after the successful completion of the previous command:

    ```
    ...
    13/04/09 08:44:51 INFO common.Storage: Storage directory
    /.../dfs/name has been successfully formatted.
    ....
    ```

7.  Start the HDFS using the following command:

    ```
    $ $HADOOP_HOME/sbin/start-dfs.sh
    ```

    This command will first start a NameNode in the master node. Then it will start the DataNode services in the machines mentioned in the `slaves` file. Finally, it'll start the secondary NameNode.

8. HDFS comes with a monitoring web console to verify the installation and to monitor the HDFS cluster. It also lets users explore the contents of the HDFS filesystem. The HDFS monitoring console can be accessed from `http://{NAMENODE}:50070/`. Visit the monitoring console and verify whether you can see the HDFS startup page. Here, replace `{NAMENODE}` with the IP address of the node running the HDFS NameNode.

9. Alternatively, you can use the following command to get a report about the HDFS status:

```
$ $HADOOP_HOME/bin/hadoop dfsadmin -report
```

10. Finally, shut down the HDFS cluster using the following command:

```
$ $HADOOP_HOME/sbin/stop-dfs.sh
```

## See also

▸ In the *HDFS command-line file operations* recipe, we will explore how to use HDFS to store and manage files.

▸ The HDFS setup is only a part of the Hadoop installation. The *Setting up Hadoop YARN in a distributed cluster environment using Hadoop v2* recipe describes how to set up the rest of Hadoop.

▸ The *Setting up Hadoop ecosystem in a distributed cluster environment using a Hadoop distribution* recipe explores how to use a packaged Hadoop distribution to install the Hadoop ecosystem in your cluster.

# Setting up Hadoop YARN in a distributed cluster environment using Hadoop v2

Hadoop v2 YARN deployment includes deploying the ResourceManager service on the master node and deploying NodeManager services in the slave nodes. YARN ResourceManager is the service that arbitrates all the resources of the cluster, and NodeManager is the service that manages the resources in a single node.

Hadoop MapReduce applications can run on YARN using a YARN ApplicationMaster to coordinate each job and a set of resource containers to run the Map and Reduce tasks.

Installing Hadoop directly using Hadoop release artifacts, as mentioned in this recipe, is recommended for development testing and for advanced use cases only. For regular production clusters, we recommend using a packaged Hadoop distribution as mentioned in the *Setting up Hadoop ecosystem in a distributed cluster environment using a Hadoop distribution* recipe. Packaged Hadoop distributions make it much easier to install, configure, maintain, and update the components of the Hadoop ecosystem.

## Getting ready

You can follow this recipe either using a single machine as a pseudo-distributed installation or using a multiple machine cluster. If you are using multiple machines, you should choose one machine as the master node where you will run the HDFS NameNode and the YARN ResourceManager. If you are using a single machine, use it as both the master node as well as the slave node.

Set up HDFS by following the *Setting up HDFS* recipe.

## How to do it...

Let's set up Hadoop YARN by setting up the YARN ResourceManager and the NodeManagers.

1. In each machine, create a directory named local inside {HADOOP_DATA_DIR}, which you created in the *Setting up HDFS* recipe. Change the directory permissions to 755.

2. Add the following to the {HADOOP_HOME}/etc/hadoop/mapred-site.xml template and save it as {HADOOP_HOME}/etc/hadoop/mapred-site.xml:

```
<property>
   <name>fs.default.name</name>
   <value>hdfs://localhost:9000</value>
</property>
```

3. Add the following to the {HADOOP_HOME}/etc/hadoop/yarn-site.xml file:

```
<property>
   <name>yarn.nodemanager.aux-services</name>
   <value>mapreduce_shuffle</value>
</property>
<property>
   <name>yarn.nodemanager.aux-services
   .mapreduce_shuffle.class</name>
   <value>org.apache.hadoop.mapred.ShuffleHandler</value>
</property>
```

4. Start HDFS using the following command:

```
$ $HADOOP_HOME/sbin/start-dfs.sh
```

5. Run the following command to start the YARN services:

```
$ $HADOOP_HOME/sbin/start-yarn.sh
starting yarn daemons
starting resourcemanager, logging to ………
xxx.xx.xxx.xxx: starting nodemanager, logging to ………
```

6. Run the following command to start the MapReduce JobHistoryServer. This enables the web console for MapReduce job histories:

```
$ $HADOOP_HOME/sbin/mr-jobhistory-daemon.sh start
historyserver
```

7. Verify the installation by listing the processes through the `jps` command. The master node will list the NameNode, ResourceManager, and JobHistoryServer services. The slave nodes will list DataNode and NodeManager services:

```
$ jps
27084 NameNode
2073 JobHistoryServer
2106 Jps
2588
1536 ResourceManager
```

8. Visit the web-based monitoring pages for ResourceManager available at `http://{MASTER_NODE}:8088/`.

## How it works...

As described in the introduction to the chapter, Hadoop v2 installation consists of HDFS nodes, YARN ResourceManager, and worker nodes. When we start the NameNode, it finds slaves through the `HADOOP_HOME/slaves` file and uses SSH to start the DataNodes in the remote server at the startup. Also, when we start ResourceManager, it finds slaves through the `HADOOP_HOME/slaves` file and starts NodeManagers.

## See also

The *Setting up Hadoop ecosystem in a distributed cluster environment using a Hadoop distribution* recipe explores how to use a packaged Hadoop distribution to install the Hadoop ecosystem in your cluster.

# Setting up Hadoop ecosystem in a distributed cluster environment using a Hadoop distribution

The Hadoop YARN ecosystem now contains many useful components providing a wide range of data processing, storing, and querying functionalities for the data stored in HDFS. However, manually installing and configuring all of these components to work together correctly using individual release artifacts is quite a challenging task. Other challenges of such an approach include the monitoring and maintenance of the cluster and the multiple Hadoop components.

Luckily, there exist several commercial software vendors that provide well integrated packaged Hadoop distributions to make it much easier to provision and maintain a Hadoop YARN ecosystem in our clusters. These distributions often come with easy GUI-based installers that guide you through the whole installation process and allow you to select and install the components that you require in your Hadoop cluster. They also provide tools to easily monitor the cluster and to perform maintenance operations. For regular production clusters, we recommend using a packaged Hadoop distribution from one of the well-known vendors to make your Hadoop journey much easier. Some of these commercial Hadoop distributions (or editions of the distribution) have licenses that allow us to use them free of charge with optional paid support agreements.

**Hortonworks Data Platform** (**HDP**) is one such well-known Hadoop YARN distribution that is available free of charge. All the components of HDP are available as free and open source software. You can download HDP from `http://hortonworks.com/hdp/downloads/`. Refer to the installation guides available in the download page for instructions on the installation.

**Cloudera CDH** is another well-known Hadoop YARN distribution. The Express edition of CDH is available free of charge. Some components of the Cloudera distribution are proprietary and available only for paying clients. You can download Cloudera Express from `http://www.cloudera.com/content/cloudera/en/products-and-services/cloudera-express.html`. Refer to the installation guides available on the download page for instructions on the installation.

Hortonworks HDP, Cloudera CDH, and some of the other vendors provide fully configured quick start virtual machine images that you can download and run on your local machine using a virtualization software product. These virtual machines are an excellent resource to learn and try the different Hadoop components as well as for evaluation purposes before deciding on a Hadoop distribution for your cluster.

Apache Bigtop is an open source project that aims to provide packaging and integration/ interoperability testing for the various Hadoop ecosystem components. Bigtop also provides a vendor neutral packaged Hadoop distribution. While it is not as sophisticated as the commercial distributions, Bigtop is easier to install and maintain than using release binaries of each of the Hadoop components. In this recipe, we provide steps to use Apache Bigtop to install Hadoop ecosystem in your local machine.

Any of the earlier mentioned distributions, including Bigtop, is suitable for the purposes of following the recipes and executing the samples provided in this book. However, when possible, we recommend using Hortonworks HDP, Cloudera CDH, or other commercial Hadoop distributions.

## Getting ready

This recipe provides instructions for the Cent OS and Red Hat operating systems. Stop any Hadoop service that you started in the previous recipes.

## How to do it...

The following steps will guide you through the installation process of a Hadoop cluster using Apache Bigtop for Cent OS and Red Hat operating systems. Please adapt the commands accordingly for other Linux-based operating systems.

1. Install the Bigtop repository:

   ```
   $ sudo wget -O \
   /etc/yum.repos.d/bigtop.repo \
   http://www.apache.org/dist/bigtop/stable/repos/centos6/bigtop.repo
   ```

2. Search for Hadoop:

   ```
   $ yum search hadoop
   ```

3. Install Hadoop v2 using Yum. This will install Hadoop v2 components (MapReduce, HDFS, and YARN) together with the ZooKeeper dependency.

   ```
   $ sudo yum install hadoop\*
   ```

4. Use your favorite editor to add the following line to the /etc/default/bigtop-utils file. It is recommended to point JAVA_HOME to a JDK 1.6 or later installation (Oracle JDK 1.7 or higher is preferred).

   ```
   export JAVA_HOME=/usr/java/default/
   ```

5. Initialize and format the NameNode:

   ```
   $ sudo  /etc/init.d/hadoop-hdfs-namenode init
   ```

6. Start the Hadoop NameNode service:

```
$ sudo service hadoop-hdfs-namenode start
```

7. Start the Hadoop DataNode service:

```
$ sudo service hadoop-hdfs-datanode start
```

8. Run the following script to create the necessary directories in HDFS:

```
$ sudo  /usr/lib/hadoop/libexec/init-hdfs.sh
```

9. Create your home directory in HDFS and apply the necessary permisions:

```
$ sudo su -s /bin/bash hdfs \
-c "/usr/bin/hdfs dfs -mkdir /user/${USER}"
$ sudo su -s /bin/bash hdfs \
-c "/usr/bin/hdfs dfs -chmod -R 755 /user/${USER}"
$ sudo su -s /bin/bash hdfs \
-c "/usr/bin/hdfs dfs -chown ${USER} /user/${USER}"
```

10. Start the YARN ResourceManager and the NodeManager:

```
$ sudo service hadoop-yarn-resourcemanager start
$ sudo service hadoop-yarn-nodemanager start
$ sudo service hadoop-mapreduce-historyserver start
```

11. Try the following commands to verify the installation:

```
$ hadoop fs -ls  /
$ hadoop jar \
/usr/lib/hadoop-mapreduce/hadoop-mapreduce-examples.jar \
pi 10 1000
```

12. You can also monitor the status of the HDFS using the monitoring console available at `http://<namenode_ip>:50070`.

13. Install Hive, HBase, Mahout, and Pig using Bigtop as follows:

```
$ sudo yum install hive\*, hbase\*, mahout\*, pig\*
```

## There's more...

▶ You can use the Puppet-based cluster installation of Bigtop by following the steps given at `https://cwiki.apache.org/confluence/display/BIGTOP/How+to+install+BigTop+0.7.0+hadoop+on+CentOS+with+puppet`

▶ You can also set up your Hadoop v2 clusters in a cloud environment as we will discuss in the next chapter

# HDFS command-line file operations

HDFS is a distributed filesystem, and just like any other filesystem, it allows users to manipulate the filesystem using shell commands. This recipe introduces some of these commands and shows how to use the HDFS shell commands.

It is worth noting that some of the HDFS commands have a one-to-one correspondence with the mostly used Unix commands. For example, consider the following command:

```
$ bin/hdfs dfs -cat /user/joe/foo.txt
```

The command reads the /user/joe/foo.txt file and prints it to the screen, just like the cat command in a Unix system.

## Getting ready

Start the HDFS server by following the *Setting up HDFS* recipe or the *Setting up Hadoop ecosystem in a distributed cluster environment using a Hadoop distribution* recipe.

## How to do it...

1. Run the following command to list the content of your HDFS home directory. If your HDFS home directory does not exist, please follow step 9 of the *Setting up Hadoop ecosystem in a distributed cluster environment using a Hadoop distribution* recipe to create your HDFS home directory.

   ```
   $ hdfs dfs -ls
   ```

2. Run the following command to create a new directory called test inside your home directory in HDFS:

   ```
   $ hdfs dfs -mkdir test
   ```

3. The HDFS filesystem has / as the root directory. Run the following command to list the content of the newly created directory in HDFS:

   ```
   $ hdfs dfs -ls test
   ```

4. Run the following command to copy the local readme file to test:

   ```
   $ hdfs dfs -copyFromLocal README.txt test
   ```

5. Run the following command to list the test directory:

   ```
   $ hdfs dfs -ls test

   Found 1 items

   -rw-r--r--   1 joesupergroup1366 2013-12-05 07:06
   /user/joe/test/README.txt
   ```

6.  Run the following command to copy the `/test/README.txt` file back to a local directory:

```
$ hdfs dfs -copyToLocal \
test/README.txt README-NEW.txt
```

## How it works...

When the command is issued, the HDFS client will talk to HDFS NameNode on our behalf and carry out the operation. The client will pick up the NameNode from the configurations in the `HADOOP_HOME/etc/hadoop/conf` directory.

However, if needed, we can use a fully qualified path to force the client to talk to a specific NameNode. For example, `hdfs://bar.foo.com:9000/data` will ask the client to talk to NameNode running on `bar.foo.com` at the port `9000`.

## There's more...

HDFS supports most of the Unix commands such as `cp`, `mv`, and `chown`, and they follow the same pattern as the commands discussed earlier. The following command lists all the available HDFS shell commands:

```
$ hdfs dfs -help
```

Using a specific command with `help` will display the usage of that command.

```
$ hdfs dfs -help du
```

# Running the WordCount program in a distributed cluster environment

This recipe describes how to run a MapReduce computation in a distributed Hadoop v2 cluster.

## Getting ready

Start the Hadoop cluster by following the *Setting up HDFS* recipe or the *Setting up Hadoop ecosystem in a distributed cluster environment using a Hadoop distribution* recipe.

## How to do it...

Now let's run the WordCount sample in the distributed Hadoop v2 setup:

1. Upload the `wc-input` directory in the source repository to the HDFS filesystem. Alternatively, you can upload any other set of text documents as well.

   ```
   $ hdfs dfs -copyFromLocal wc-input .
   ```

2. Execute the WordCount example from the HADOOP_HOME directory:

   ```
   $ hadoop jar hcb-c1-samples.jar \
   chapter1.WordCount \
   wc-input wc-output
   ```

3. Run the following commands to list the output directory and then look at the results:

   ```
   $hdfs dfs -ls wc-output

   Found 3 items

   -rw-r--r--    1 joesupergroup0 2013-11-09 09:04
   /data/output1/_SUCCESS

   drwxr-xr-x    - joesupergroup0 2013-11-09 09:04
   /data/output1/_logs

   -rw-r--r--    1 joesupergroup1306 2013-11-09 09:04
   /data/output1/part-r-00000


   $ hdfs dfs -cat wc-output/part*
   ```

## How it works...

When we submit a job, YARN would schedule a MapReduce ApplicationMaster to coordinate and execute the computation. ApplicationMaster requests the necessary resources from the ResourceManager and executes the MapReduce computation using the containers it received from the resource request.

## There's more...

You can also see the results of the WordCount application through the HDFS monitoring UI by visiting http://NAMANODE:50070.

# Benchmarking HDFS using DFSIO

Hadoop contains several benchmarks that you can use to verify whether your HDFS cluster is set up properly and performs as expected. DFSIO is a benchmark test that comes with Hadoop, which can be used to analyze the I/O performance of an HDFS cluster. This recipe shows how to use DFSIO to benchmark the read/write performance of an HDFS cluster.

## Getting ready

You must set up and deploy HDFS and Hadoop v2 YARN MapReduce prior to running these benchmarks. Locate the `hadoop-mapreduce-client-jobclient-*-tests.jar` file in your Hadoop installation.

## How to do it...

The following steps will show you how to run the write and read DFSIO performance benchmarks:

1. Execute the following command to run the HDFS write performance benchmark. The `-nrFiles` parameter specifies the number of files to be written by the benchmark. Use a number high enough to saturate the task slots in your cluster. The `-fileSize` parameter specifies the file size of each file in MB. Change the location of the `hadoop-mapreduce-client-jobclient-*-tests.jar` file in the following commands according to your Hadoop installation.

```
$ hadoop jar \
$HADOOP_HOME/share/hadoop/mapreduce/hadoop-
mapreduce-client-jobclient-*-tests.jar \
TestDFSIO -write -nrFiles 32 -fileSize 1000
```

2. The write benchmark writes the results to the console as well as appending to a file named `TestDFSIO_results.log`. You can provide your own result filename using the `-resFile` parameter.

3. The following step will show you how to run the HDFS read performance benchmark. The read performance benchmark uses the files written by the write benchmark in step 1. Hence, the write benchmark should be executed before running the read benchmark and the files written by the write benchmark should exist in the HDFS for the read benchmark to work properly. The benchmark writes the results to the console and appends the results to a logfile similarly to the write benchmark.

```
$hadoop jar \
$HADOOP_HOME/share/Hadoop/mapreduce/hadoop-
mapreduce-client-jobclient-*-tests.jar \
TestDFSIO -read \
-nrFiles 32 -fileSize 1000
```

4. The files generated by the preceding benchmarks can be cleaned up using the following command:

```
$hadoop jar \
$HADOOP_HOME/share/Hadoop/mapreduce/hadoop-
mapreduce-client-jobclient-*-tests.jar \
TestDFSIO -clean
```

## How it works...

DFSIO executes a MapReduce job where the Map tasks write and read the files in parallel, while the Reduce tasks are used to collect and summarize the performance numbers. You can compare the throughput and IO rate results of this benchmark with the total number of disks and their raw speeds to verify whether you are getting the expected performance from your cluster. Please note the replication factor when verifying the write performance results. High standard deviation in these tests may hint at one or more underperforming nodes due to some reason.

## There's more...

Running these tests together with monitoring systems can help you identify the bottlenecks of your Hadoop cluster much easily.

# Benchmarking Hadoop MapReduce using TeraSort

Hadoop TeraSort is a well-known benchmark that aims to sort 1 TB of data as fast as possible using Hadoop MapReduce. TeraSort benchmark stresses almost every part of the Hadoop MapReduce framework as well as the HDFS filesystem making it an ideal choice to fine-tune the configuration of a Hadoop cluster.

The original TeraSort benchmark sorts 10 million 100 byte records making the total data size 1 TB. However, we can specify the number of records, making it possible to configure the total size of data.

## Getting ready

You must set up and deploy HDFS and Hadoop v2 YARN MapReduce prior to running these benchmarks, and locate the hadoop-mapreduce-examples-*.jar file in your Hadoop installation.

## How to do it...

The following steps will show you how to run the TeraSort benchmark on the Hadoop cluster:

1.  The first step of the TeraSort benchmark is the data generation. You can use the `teragen` command to generate the input data for the TeraSort benchmark. The first parameter of `teragen` is the number of records and the second parameter is the HDFS directory to generate the data. The following command generates 1 GB of data consisting of 10 million records to the `tera-in` directory in HDFS. Change the location of the `hadoop-mapreduce-examples-*.jar` file in the following commands according to your Hadoop installation:

    ```
    $ hadoop jar \
    $HADOOP_HOME/share/Hadoop/mapreduce/hadoop-
    mapreduce-examples-*.jar \
    teragen 10000000 tera-in
    ```

> It's a good idea to specify the number of Map tasks to the `teragen` computation to speed up the data generation. This can be done by specifying the `–Dmapred.map.tasks` parameter.
>
> Also, you can increase the HDFS block size for the generated data so that the Map tasks of the TeraSort computation would be coarser grained (the number of Map tasks for a Hadoop computation typically equals the number of input data blocks). This can be done by specifying the `–Ddfs.block. size` parameter.
>
> ```
> $ hadoop jar $HADOOP_HOME/share/hadoop/mapreduce/hadoop-
> mapreduce-examples-*.jar \
> teragen –Ddfs.block.size=536870912 \
> –Dmapred.map.tasks=256 10000000 tera-in
> ```

2.  The second step of the TeraSort benchmark is the execution of the TeraSort MapReduce computation on the data generated in step 1 using the following command. The first parameter of the `terasort` command is the input of HDFS data directory, and the second part of the `terasort` command is the output of the HDFS data directory.

    ```
    $ hadoop jar \
    $HADOOP_HOME/share/hadoop/mapreduce/hadoop-
    mapreduce-examples-*.jar \
    terasort tera-in tera-out
    ```

It's a good idea to specify the number of Reduce tasks to the TeraSort computation to speed up the Reducer part of the computation. This can be done by specifying the `-Dmapred.reduce.tasks` parameter as follows:

```
$ hadoop jar $HADOOP_HOME/share/hadoop/mapreduce/hadoop-
mapreduce-examples-*.jar terasort -Dmapred.reduce.
tasks=32 tera-in tera-out
```

3.  The last step of the TeraSort benchmark is the validation of the results. This can be done using the `teravalidate` application as follows. The first parameter is the directory with the sorted data and the second parameter is the directory to store the report containing the results.

```
$ hadoop jar \
$HADOOP_HOME/share/hadoop/mapreduce/hadoop-
mapreduce-examples-*.jar \
teravalidate tera-out tera-validate
```

## How it works...

TeraSort uses the sorting capability of the MapReduce framework together with a custom range **Partitioner** to divide the Map output among the Reduce tasks ensuring the global sorted order.

# 2

# Cloud Deployments – Using Hadoop YARN on Cloud Environments

In this chapter, we will cover the following recipes:

- ▶ Running Hadoop MapReduce v2 computations using Amazon Elastic MapReduce
- ▶ Saving money using Amazon EC2 Spot Instances to execute EMR job flows
- ▶ Executing a Pig script using EMR
- ▶ Executing a Hive script using EMR
- ▶ Creating an Amazon EMR job flow using the AWS Command Line Interface
- ▶ Deploying an Apache HBase cluster on Amazon EC2 using EMR
- ▶ Using EMR bootstrap actions to configure VMs for the Amazon EMR jobs
- ▶ Using Apache Whirr to deploy an Apache Hadoop cluster in EC2 environment

## Introduction

In this chapter, we will explore several mechanisms to deploy and execute Hadoop MapReduce v2 and other Hadoop-related computations on cloud environments.

Cloud computing environments such as Amazon EC2 and Microsoft Azure provide on-demand compute and storage resources as a service over the Web. These cloud computing environments enable us to perform occasional large-scale Hadoop computations without an upfront capital investment and require us to pay only for the actual usage. Another advantage of using cloud environments is the ability to increase the throughput of the Hadoop computations by horizontally scaling the number of computing resources with minimal additional cost. For an example, the cost for using 10 cloud instances for 100 hours equals the cost of using 100 cloud instances for 10 hours. In addition to storage, compute, and hosted MapReduce services, these cloud environments provide many other distributed computing services as well, which you may find useful when implementing your overall application architecture.

While the cloud environments provide many advantages over their traditional counterparts, they also come with several unique reliability and performance challenges due to the virtualized, multi-tenant nature of the infrastructure. With respect to the data-intensive Hadoop computations, one of the major challenges would be the transfer of large datasets in and out of the cloud environments. We also need to make sure to use a persistent storage medium to store any data that you need to preserve. Any data that is stored in the ephemeral instance storage of the cloud instances would be lost at the termination of those instances.

We will mainly be using the Amazon AWS cloud for the recipes in this chapter due to the maturity of the Linux instance support and the maturity of hosted Hadoop services compared to the other commercial cloud offerings such as Microsoft Azure cloud.

This chapter guides you on using Amazon **Elastic MapReduce** (**EMR**), which is the hosted Hadoop infrastructure, to execute traditional MapReduce computations as well as Pig and Hive computations on the Amazon EC2 infrastructure. This chapter also presents how to provision an HBase cluster using Amazon EMR and how to back up and restore the data of an EMR HBase cluster. We will also use Apache Whirr, a cloud neutral library for deploying services on cloud environments, to provision Apache Hadoop and Apache HBase clusters on cloud environments.

**Sample code**

The example code files for this book are available in the `https://github.com/thilg/hcb-v2` repository. The `chapter2` folder of the code repository contains the sample source code files for this chapter.

# Running Hadoop MapReduce v2 computations using Amazon Elastic MapReduce

**Amazon Elastic MapReduce** (**EMR**) provides on-demand managed Hadoop clusters in the **Amazon Web Services** (**AWS**) cloud to perform your Hadoop MapReduce computations. EMR uses Amazon **Elastic Compute Cloud** (**EC2**) instances as the compute resources. EMR supports reading input data from Amazon **Simple Storage Service** (**S3**) and storing of the output data in Amazon S3 as well. EMR takes care of the provisioning of cloud instances, configuring the Hadoop cluster, and the execution of our MapReduce computational flows.

In this recipe, we are going to execute the WordCount MapReduce sample (the *Writing a WordCount MapReduce application, bundling it, and running it using the Hadoop local mode* recipe from *Chapter 1, Getting Started with Hadoop v2*) in the Amazon EC2 using the Amazon Elastic MapReduce service.

## Getting ready

Build the `hcb-c1-samples.jar` file by running the Gradle build in the `chapter1` folder of the sample code repository.

## How to do it...

The following are the steps for executing WordCount MapReduce application on Amazon Elastic MapReduce:

1. Sign up for an AWS account by visiting `http://aws.amazon.com`.

2. Open the Amazon S3 monitoring console at `https://console.aws.amazon.com/s3` and sign in.

3. Create an **S3 bucket** to upload the input data by clicking on **Create Bucket**. Provide a unique name for your bucket. Let's assume the name of the bucket is `wc-input-data`. You can find more information on creating an S3 bucket at `http://docs.amazonwebservices.com/AmazonS3/latest/gsg/CreatingABucket.html`. There also exist several third-party desktop clients for the Amazon S3. You can use one of those clients to manage your data in S3 as well.

4.  Upload your input data to the bucket we just created by selecting the bucket and clicking on **Upload**. The input data for the WordCount sample should be one or more text files:

5.  Create an S3 bucket to upload the JAR file needed for our MapReduce computation. Let's assume the name of the bucket as `sample-jars`. Upload `hcb-c1-samples.jar` to the newly created bucket.

6.  Create an S3 bucket to store the output data of the computation. Let's assume the name of this bucket as `wc-output-data`. Create another S3 bucket to store the logs of the computation. Let's assume the name of this bucket is `hcb-c2-logs`.

 Note that all the S3 users share the S3 bucket naming namespace. Hence, using the example bucket names given in this recipe might not work for you. In such scenarios, you should give your own custom names for the buckets and substitute those names in the subsequent steps of this recipe.

7.  Open the Amazon EMR console at `https://console.aws.amazon.com/elasticmapreduce`. Click on the **Create Cluster** button to create a new EMR cluster. Provide a name for your cluster.

8.  In the **Log folder S3 location** textbox, enter the path of the S3 bucket you created earlier to store the logs. Select the **Enabled** radio button for **Debugging**.

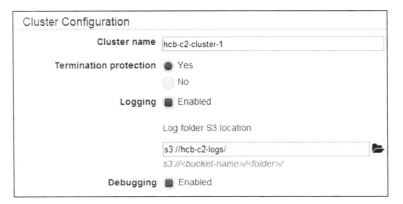

9. Select the Hadoop distribution and version in the **Software Configuration** section. Select AMI version 3.0.3 or above with the Amazon Hadoop distribution to deploy a Hadoop v2 cluster. Leave the default selected applications (Hive, Pig, and Hue) in the **Application to be installed** section.

10. Select the EC2 instance types, instance counts, and the availability zone in the **Hardware Configuration** section. The default options use two EC2 m1.large instances for the Hadoop slave nodes and one EC2 m1.large instance for the Hadoop Master node.

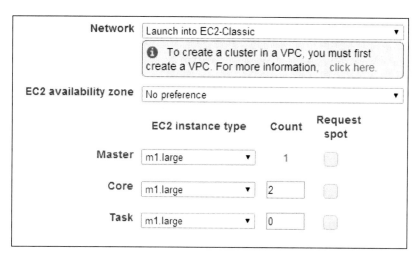

11. Leave the default options in the **Security and Access** and **Bootstrap Actions** sections.

12. Select the **Custom Jar** option under the **Add Step** dropdown of the **Steps** section. Click on **Configure and add** to configure the JAR file for our computation. Specify the S3 location of `hcb-c1-samples.jar` in the **Jar S3 location** textbox. You should specify the location of the JAR in the format `s3n://bucket_name/jar_name`. In the **Arguments** textbox, type `chapter1.WordCount` followed by the bucket location where you uploaded the input data in step 4 and the output data bucket you created in step 6. The output path should not exist and we use a directory (for example, `wc-output-data/out1`) inside the output bucket you created in step 6 as the output path. You should specify the locations using the format, `s3n://bucket_name/path`.

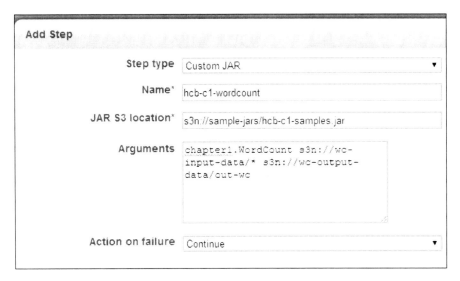

13. Click on **Create Cluster** to launch the EMR Hadoop cluster and run the WordCount application.

Amazon will charge you for the compute and storage resources you use when clicking on **Create Cluster** in step 13. Refer to the *Saving money using Amazon EC2 Spot Instances to execute EMR job flows* recipe to find out how you can save money by using Amazon EC2 Spot Instances.

Note that AWS bills you by the hour and any partial usage would get billed as an hour. Each launch and stop of an instance would be billed as a single hour, even if it takes only minutes. Be aware of the expenses when performing frequent re-launching of clusters for testing purposes.

14. Monitor the progress of your MapReduce cluster deployment and the computation in the **Cluster List | Cluster Details** page of the Elastic MapReduce console. Expand the **Steps** section of the page to see the status of the individual steps of the cluster setup and the application execution. Select a step and click on **View logs** to view the logs and to debug the computation. Since EMR uploads the logfiles periodically, you might have to wait and refresh to access the logfiles. Check the output of the computation in the output data bucket using the AWS S3 console.

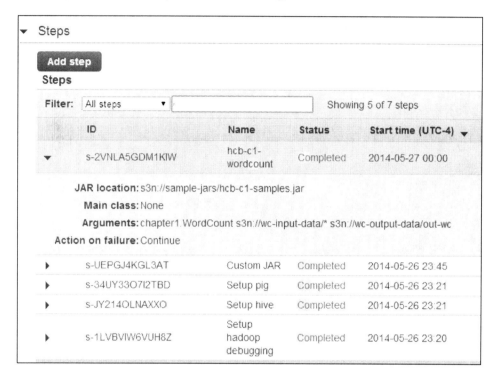

15. Terminate your cluster to avoid getting billed for the instances that are left. However, you may leave the cluster running to try out the other recipes in this chapter.

## See also

▸ The *Writing a WordCount MapReduce application, bundling it, and running it using the Hadoop local mode* recipe from *Chapter 1, Getting Started with Hadoop v2*

▸ The *Running the WordCount program in a distributed cluster environment* recipe from *Chapter 1, Getting Started with Hadoop v2*

# Saving money using Amazon EC2 Spot Instances to execute EMR job flows

Amazon **EC2 Spot Instances** allow us to purchase underutilized EC2 compute resources at a significant discount. The prices of Spot Instances change depending on the demand. We can submit bids for the Spot Instances and we receive the requested compute instances if our bid exceeds the current Spot Instance price. Amazon bills these instances based on the actual Spot Instance price, which can be lower than your bid. Amazon will terminate your instances, if the Spot Instance price exceeds your bid. However, Amazon does not charge for partial Spot Instance hours if Amazon terminated your instances. You can find more information on Amazon EC2 Spot Instances at `http://aws.amazon.com/ec2/spot-instances/`.

Amazon EMR supports using Spot Instances both as master as well as worker compute instances. Spot Instances are ideal to execute nontime critical computations such as batch jobs.

## How to do it...

The following steps show you how to use Amazon EC2 Spot Instances with Amazon Elastic MapReduce to execute the WordCount MapReduce application:

1.  Follow steps 1 to 9 of the *Running Hadoop MapReduce v2 computations using Amazon Elastic MapReduce* recipe.

2.  Configure your EMR cluster to use Spot Instances in the **Hardware Configuration** section. (Refer to step 10 of the *Running Hadoop MapReduce v2 computations using Amazon Elastic MapReduce* recipe). In the **Hardware Configuration** section, select the **Request Spot** checkboxes next to the instance types.

3. Specify your bid price in the **Bid price** textboxes. You can find the Spot Instance pricing history in the Spot Requests window of the Amazon EC2 console (`https://console.aws.amazon.com/ec2`).

Hardware Configuration

❶ Specify the networking and hardware configuration for your cluster. If you need more than 20 EC2 instances, complete this form. Request Spot instances (unused EC2 capacity) to save money.

| | | | | | |
|---|---|---|---|---|---|
| Network | Launch into EC2-Classic ▼ | | | | Use a Virtual Private Cloud (VPC) to process sensitive data or connect to a private network.   Create a VPC |
| | ❶ To create a cluster in a VPC, you must first create a VPC. For more information,   click here. | | | | |
| EC2 availability zone | No preference ▼ | | | | Launch the cluster in a specific EC2 Availability Zone. |

| | EC2 instance type | Count | Request spot | Bid price | |
|---|---|---|---|---|---|
| Master | m1.large ▼ | 1 | ⬛ | 0.1 | The Master instance assigns Hadoop tasks to core and task nodes, and monitors their status. |
| Core | m1.large ▼ | 2 | ⬛ | 0.1 | Core instances run Hadoop tasks and store data using the Hadoop Distributed File System (HDFS). |
| Task | m1.large ▼ | 0 | ⬜ | | Task instances run Hadoop tasks. |

4. Follow steps 11 to 16 of the *Running Hadoop MapReduce v2 computations using Amazon Elastic MapReduce* recipe.

## There's more...

You can also run the EMR computations on a combination of traditional EC2 on-demand instances and EC2 Spot Instances, safe guarding your computation against possible Spot Instance terminations.

Since Amazon bills the Spot Instances using the current Spot price irrespective of your bid price, it is a good practice to not set the Spot Instance price too low to avoid the risk of frequent terminations.

## See also

The *Running Hadoop MapReduce v2 computations using Amazon Elastic MapReduce* recipe.

# Executing a Pig script using EMR

Amazon EMR supports executing Apache Pig scripts on the data stored in S3. Refer to the Pig-related recipes in *Chapter 7, Hadoop Ecosystem II – Pig, HBase, Mahout, and Sqoop,* for more details on using Apache Pig for data analysis.

In this recipe, we are going to execute a simple Pig script using Amazon EMR. This sample will use the Human Development Reports data (`http://hdr.undp.org/en/statistics/data/`) to print names of countries that have a GNI value greater than $2000 of gross national income per capita (GNI) sorted by GNI.

## How to do it...

The following steps show you how to use a Pig script with Amazon Elastic MapReduce to process a dataset stored on Amazon S3:

1.  Use the Amazon S3 console to create a bucket in S3 to upload the input data. Upload the `resources/hdi-data.csv` file in the source repository for this chapter to the newly created bucket. You can also use an existing bucket or a directory inside a bucket as well. We assume the S3 path for the uploaded file is `hcb-c2-data/hdi-data.csv`.

2.  Review the Pig script available in the `resources/countryFilter-EMR.pig` file of the source repository for this chapter. This script uses the STORE command to save the result in the filesystem. In addition, we parameterize the LOAD command of the Pig script by adding `$INPUT` as the input file and the store command by adding `$OUTPUT` as the output directory. These two parameters would be substituted by the S3 input and output locations we specify in step 5.

    ```
    A = LOAD '$INPUT' using PigStorage(',')  AS
    (id:int, country:chararray, hdi:float, lifeex:int,
    mysch:int, eysch:int, gni:int);
    B = FILTER A BY gni > 2000;
    C = ORDER B BY gni;
    STORE C into '$OUTPUT';
    ```

3.  Use the Amazon S3 console to create a bucket in S3 to upload the Pig script. Upload the `resources/countryFilter-EMR.pig` script to the newly created bucket. You can also use an existing bucket or a directory inside a bucket as well. We assume the S3 path for the uploaded file as `hcb-c2-resources/countryFilter-EMR.pig`.

4.  Open the Amazon EMR console at `https://console.aws.amazon.com/elasticmapreduce`. Click on the **Create Cluster** button to create a new EMR cluster. Provide a name for your cluster. Follow steps 8 to 11 of the *Running Hadoop MapReduce v2 computations using Amazon Elastic MapReduce* recipe to configure your cluster.

You can reuse the EMR cluster you created in the *Running Hadoop MapReduce v2 computations using Amazon Elastic MapReduce* recipe to follow the steps of this recipe. To do that, use the **Add Step** option in the **Cluster Details** page of the running cluster to perform the actions mentioned in step 5.

5.  Select the **Pig Program** option under the **Add Step** dropdown of the **Steps** section. Click on **Configure and add** to configure the Pig script, input, and output data for our computation. Specify the S3 location of the Pig script we uploaded in step 3, in the **Script S3 location** textbox. You should specify the location of the script in the format `s3://bucket_name/script_filename`. Specify the S3 location of the uploaded input data file in the **Input S3 Location** textbox. In the **Output S3 Location** textbox, specify an S3 location to store the output. The output path should not exist; we use a non-existing directory (for example, `hcb-c2-out/pig`) inside the output bucket as the output path. You should specify the locations using the format `s3://bucket_name/path`. Click on **Add**.

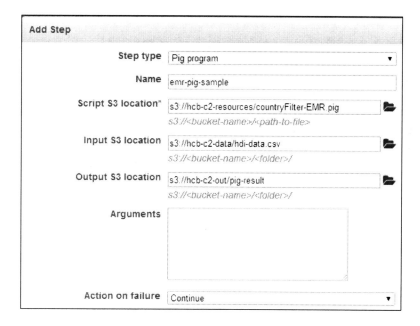

6.  Click on **Create Cluster** to launch the EMR Hadoop cluster and to run the configured Pig script.

Amazon will charge you for the compute and storage resources you use by clicking on **Create Job Flow** in step 8. Refer to the *Saving money using EC2 Spot Instances to execute EMR job flows* recipe that we discussed earlier to find out how you can save money by using Amazon EC2 Spot Instances.

7. Monitor the progress of your MapReduce cluster deployment and the computation in the **Cluster List | Cluster Details** page of the Elastic MapReduce console. Expand and refresh the **Steps** section of the page to see the status of the individual steps of the cluster setup and the application execution. Select a step and click on **View logs** to view the logs and to debug the computation. Check the output of the computation in the output data bucket using the AWS S3 console.

## There's more...

Amazon EMR allows us to use Apache Pig in the interactive mode as well.

### Starting a Pig interactive session

1. Open the Amazon EMR console at `https://console.aws.amazon.com/elasticmapreduce`. Click on the **Create Cluster** button to create a new EMR cluster. Provide a name for your cluster.

2. You must select a key pair from the **Amazon EC2 Key Pair** dropdown in the **Security and Access** section. If you do not have a usable Amazon EC2 key pair with access to the private key, log on to the Amazon EC2 console and create a new key pair.

3. Click on **Create Cluster** without specifying any steps. Make sure **No** is selected in the **Auto-Terminate** option under the **Steps** section.

4. Monitor the progress of your MapReduce cluster deployment and the computation in the **Cluster Details** page under **Cluster List** of the Elastic MapReduce console. Retrieve **Master Public DNS** from the cluster details in this page.

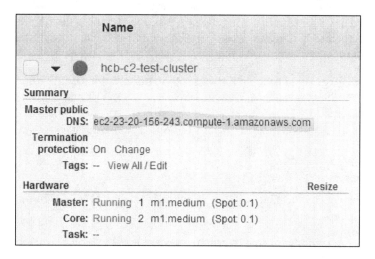

5. Use the master public DNS name and the private key file of the Amazon EC2 key pair you specified in step 2 to SSH into the master node of the cluster:

```
$ ssh -i <path-to-the-key-file> hadoop@<master-public-DNS>
```

6. Start the Pig interactive Grunt shell in the master node and issue your Pig commands:

```
$ pig
.........
grunt>
```

# Executing a Hive script using EMR

Hive provides a SQL-like query layer for the data stored in HDFS utilizing Hadoop MapReduce underneath. Amazon EMR supports executing Hive queries on the data stored in S3. Refer to the Apache Hive recipes in *Chapter 6, Hadoop Ecosystem – Apache Hive*, for more information on using Hive for large-scale data analysis.

In this recipe, we are going to execute a Hive script to perform the computation we did in the *Executing a Pig script using EMR* recipe earlier. We will use the Human Development Reports data (http://hdr.undp.org/en/statistics/data/) to print names of countries that have a GNI value greater than $2000 of gross national income per capita (GNI) sorted by GNI.

## How to do it...

The following steps show how to use a Hive script with Amazon Elastic MapReduce to query a dataset stored on Amazon S3:

1. Use the Amazon S3 console to create a bucket in S3 to upload the input data. Create a directory inside the bucket. Upload the `resources/hdi-data.csv` file in the source package of this chapter to the newly created directory inside the bucket. You can also use an existing bucket or a directory inside a bucket as well. We assume the S3 path for the uploaded file is `hcb-c2-data/data/hdi-data.csv`.

2. Review the Hive script available in the `resources/countryFilter-EMR.hql` file of the source repository for this chapter. This script first creates a mapping of the input data to a Hive table. Then we create a Hive table to store the results of our query. Finally, we issue a query to select the list of countries with a GNI larger than $2000. We use the `$INPUT` and `$OUTPUT` variables to specify the location of the input data and the location to store the output table data.

```
CREATE EXTERNAL TABLE
hdi(
    id INT,
    country STRING,
    hdi FLOAT,
    lifeex INT,
    mysch INT,
    eysch INT,
    gni INT)
ROW FORMAT DELIMITED
FIELDS TERMINATED BY ','
```

```
STORED AS TEXTFILE
LOCATION '${INPUT}';

CREATE EXTERNAL TABLE
output_countries(
    country STRING,
    gni INT)
    ROW FORMAT DELIMITED
    FIELDS TERMINATED BY ','
    STORED AS TEXTFILE
    LOCATION '${OUTPUT}';

INSERT OVERWRITE TABLE
output_countries
  SELECT
    country, gni
  FROM
    hdi
  WHERE
    gni > 2000;
```

3.  Use the Amazon S3 console to create a bucket in S3 to upload the Hive script. Upload the `resources/countryFilter-EMR.hql` script to the newly created bucket. You can also use an existing bucket or a directory inside a bucket as well. We assume the S3 path for the uploaded file is `hcb-resources/countryFilter-EMR.hql`.

4.  Open the Amazon EMR console at `https://console.aws.amazon.com/elasticmapreduce`. Click on the **Create Cluster** button to create a new EMR cluster. Provide a name for your cluster. Follow steps 8 to 11 of the *Running Hadoop MapReduce v2 computations using Amazon Elastic MapReduce* recipe to configure your cluster.

> You can reuse an EMR cluster you created for one of the earlier recipes to follow the steps of this recipe. To do that, use the **Add Step** option in the **Cluster Details** page of the running cluster to perform the actions mentioned in step 5.

5. Select the **Hive Program** option under the **Add Step** dropdown of the **Steps** section. Click on **Configure and add** to configure the Hive script, and input and output data for our computation. Specify the S3 location of the Hive script we uploaded in step 3 in the **Script S3 location** textbox. You should specify the location of the script in the format `s3://bucket_name/script_filename`. Specify the S3 location of the uploaded input data directory in the **Input S3 Location** textbox. In the **Output S3 Location** textbox, specify an S3 location to store the output. The output path should not exist and we use a nonexisting directory (for example, `hcb-c2-out/hive`) inside the output bucket as the output path. You should specify the locations using the format `s3://bucket_name/path`. Click on **Add**.

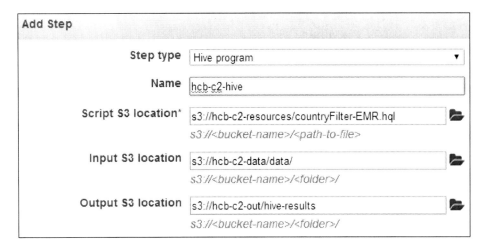

6. Click on **Create Cluster** to launch the EMR Hadoop cluster and to run the configured Hive script.

> Amazon will charge you for the compute and storage resources you use by clicking on **Create Job Flow** in step 8. Refer to the *Saving money using Amazon EC2 Spot Instances to execute EMR job flows to execute EMR job flows* recipe that we discussed earlier to find out how you can save money by using Amazon EC2 Spot Instances.

7. Monitor the progress of your MapReduce cluster deployment and the computation in the **Cluster Details** page under **Cluster List** of the Elastic MapReduce console. Expand and refresh the **Steps** section of the page to see the status of the individual steps of the cluster setup and the application execution. Select a step and click on **View logs** to view the logs and to debug the computation. Check the output of the computation in the output data bucket using the AWS S3 console.

## There's more...

Amazon EMR also allows us to use the Hive shell in the interactive mode as well.

### Starting a Hive interactive session

Follow steps 1 to 5 of the *Starting a Pig interactive session* section of the previous *Executing a Pig script using EMR* recipe to create a cluster and to log in to it using SSH.

Start the Hive shell in the master node and issue your Hive queries:

```
$ hive
hive >
.........
```

## See also

The *Simple SQL-style data querying using Apache Hive* recipe of *Chapter 6, Hadoop Ecosystem – Apache Hive*.

# Creating an Amazon EMR job flow using the AWS Command Line Interface

AWS **Command Line Interface** (**CLI**) is a tool that allows us to manage our AWS services from the command line. In this recipe, we use AWS CLI to manage Amazon EMR services.

This recipe creates an EMR job flow using the AWS CLI to execute the WordCount sample from the *Running Hadoop MapReduce computations using Amazon Elastic MapReduce* recipe of this chapter.

## Getting ready

The following are the prerequisites to get started with this recipe:

- ▸ Python 2.6.3 or higher
- ▸ pip—Python package management system

## How to do it...

The following steps show you how to create an EMR job flow using the EMR command-line interface:

1. Install AWS CLI in your machine using the pip installer:

   ```
   $ sudo pip install awscli
   ```

    Refer to `http://docs.aws.amazon.com/cli/latest/userguide/installing.html` for more information on installing the AWS CLI. This guide provides instructions on installing AWS CLI without `sudo` as well as instructions on installing AWS CLI using alternate methods.

2. Create an access key ID and a secret access key by logging in to the AWS IAM console (`https://console.aws.amazon.com/iam`). Download and save the key file in a safe location.

3. Use the `aws configure` utility to configure your AWS account to the AWC CLI. Provide the access key ID and the secret access key you obtained in the previous step. This information would get stored in the `.aws/config` and `.aws/credentials` files in your home directory.

   ```
   $ aws configure
   AWS Access Key ID [None]: AKIA….
   AWS Secret Access Key [None]: GC…
   Default region name [None]: us-east-1a
   Default output format [None]:
   ```

    You can skip to step 7 if you have completed steps 2 to 6 of the *Running Hadoop MapReduce computations using Amazon Elastic MapReduce* recipe in this chapter.

4. Create a bucket to upload the input data by clicking on **Create Bucket** in the Amazon S3 monitoring console (`https://console.aws.amazon.com/s3`). Provide a unique name for your bucket. Upload your input data to the newly-created bucket by selecting the bucket and clicking on **Upload**. The input data for the WordCount sample should be one or more text files.

5. Create an S3 bucket to upload the JAR file needed for our MapReduce computation. Upload `hcb-c1-samples.jar` to the newly created bucket.

6. Create an S3 bucket to store the output data of the computation. Create another S3 bucket to store the logs of the computation. Let's assume the name of this bucket is `hcb-c2-logs`.

7. Create an EMR cluster by executing the following command. This command will output the cluster ID of the created EMR cluster:

```
$ aws emr create-cluster --ami-version 3.1.0 \
--log-uri s3://hcb-c2-logs \
--instance-groups \
InstanceGroupType=MASTER,InstanceCount=1,\
InstanceType=m3.xlarge \
InstanceGroupType=CORE,InstanceCount=2,\
InstanceType=m3.xlarge
{
     "ClusterId": "j-2X9TDN6T041ZZ"
}
```

8. You can use the `list-clusters` command to check the status of the created EMR cluster:

```
$ aws emr list-clusters
{
    "Clusters": [
        {
            "Status": {
                "Timeline": {
                    "ReadyDateTime": 1421128629.1830001,
                    "CreationDateTime": 1421128354.4130001
                },
                "State": "WAITING",
                "StateChangeReason": {
                    "Message": "Waiting after step completed"
                }
            },
            "NormalizedInstanceHours": 24,
            "Id": "j-2X9TDN6T041ZZ",
            "Name": "Development Cluster"
        }
    ]
}
```

9. Add a job step to this EMR cluster by executing the following. Replace the paths of the JAR file, input data location, and the output data location with the locations you used in steps 5, 6, and 7. Replace `cluster-id` with the cluster ID of your newly created EMR cluster.

```
$ aws emr add-steps \
--cluster-id j-2X9TDN6T041ZZ \
--steps Type=CUSTOM_JAR,Name=CustomJAR,ActionOnFailure=CONTINUE,\
Jar=s3n://[S3 jar file bucket]/hcb-c1-samples.jar,\
Args=chapter1.WordCount,\
s3n://[S3 input data path]/*,\
s3n://[S3 output data path]/wc-out
{
    "StepIds": [
        "s-1SEEPDZ99H3Y2"
    ]
}
```

10. Check the status of the submitted job step using the `describe-step` command as follows. You can also check the status and debug your job flow using the Amazon EMR console (`https://console.aws.amazon.com/elasticmapreduce`).

```
$ aws emr describe-step \
-cluster-id j-2X9TDN6T041ZZ \
-step-id s-1SEEPDZ99H3Y2
```

11. Once the job flow is completed, check the result of the computation in the output data location using the S3 console.

12. Terminate the cluster using the `terminate-clusters` command:

```
$ aws emr terminate-clusters --cluster-ids j-2X9TDN6T041ZZ
```

## There's more...

You can use EC2 Spot Instances with your EMR clusters to reduce the cost of your computations. Add a bid price to your request by adding the `--BidPrice` parameter to the instance groups of your `create-cluster` command:

```
$ aws emr create-cluster --ami-version 3.1.0 \
--log-uri s3://hcb-c2-logs \
--instance-groups \
InstanceGroupType=MASTER,InstanceCount=1,\
InstanceType=m3.xlarge,BidPrice=0.10 \
InstanceGroupType=CORE,InstanceCount=2,\
InstanceType=m3.xlarge,BidPrice=0.10
```

Refer to the *Saving money using Amazon EC2 Spot Instances to execute EMR job flows* recipe in this chapter for more details on Amazon Spot Instances.

## See also

▸ The *Running Hadoop MapReduce computations using Amazon Elastic MapReduce* recipe of this chapter

▸ You can find the reference documentation for the EMR section of the AWS CLI at `http://docs.aws.amazon.com/cli/latest/reference/emr/index.html`

# Deploying an Apache HBase cluster on Amazon EC2 using EMR

We can use Amazon Elastic MapReduce to start an Apache HBase cluster on the Amazon infrastructure to store large quantities of data in a column-oriented data store. We can use the data stored on Amazon EMR HBase clusters as input and output of EMR MapReduce computations as well. We can incrementally back up the data stored in Amazon EMR HBase clusters to Amazon S3 for data persistence. We can also start an EMR HBase cluster by restoring the data from a previous S3 backup.

In this recipe, we start an Apache HBase cluster on Amazon EC2 using Amazon EMR; perform several simple operations on the newly created HBase cluster and back up the HBase data into Amazon S3 before shutting down the cluster. Then we start a new HBase cluster restoring the HBase data backups from the original HBase cluster.

## Getting ready

You should have the AWS CLI installed and configured to manually back up HBase data. Refer to the *Creating an Amazon EMR job flow using the AWS Command Line Interface* recipe in this chapter for more information on installing and configuring the AWS CLI.

## How to do it...

The following steps show how to deploy an Apache HBase cluster on Amazon EC2 using Amazon EMR:

1. Create an S3 bucket to store the HBase backups. We assume the S3 bucket for the HBase data backups is `hcb-c2-data`.

2. Open the Amazon EMR console at `https://console.aws.amazon.com/elasticmapreduce`. Click on the **Create Cluster** button to create a new EMR cluster. Provide a name for your cluster.

3. Provide a path in **Log folder S3 location** and select an **AMI version** with Hadoop v2 (for example, AMI version 3.1.0 with Hadoop 2.4.0).

4. Select **HBase** from the **Additional Applications** drop-down box under the **Applications to be installed** section. Click on **Configure and add**.

5. Make sure the **Restore from backup** radio button is not selected. Select the **Schedule regular backups** and **Consistent Backup** radio buttons. Specify a **Backup frequency** for automatic scheduled incremental data backups and provide a path inside the Blob we created in step 1 as the backup location. Click on **Continue**.

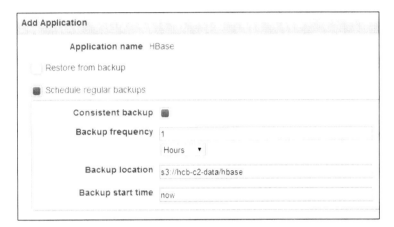

6. Configure the EC2 instances under the **Hardware Configuration** section.

7. Select a key pair in the **Amazon EC2 Key Pair** drop-down box. Make sure you have the private key for the selected EC2 key pair downloaded on your computer.

If you do not have a usable key pair, go to the EC2 console (`https://console.aws.amazon.com/ec2`) to create a key pair. To create a key pair, log in to the EC2 dashboard, select a region, and click on **Key Pairs** under the **Network and Security** menu. Click on the **Create Key Pair** button in the **Key Pairs** window and provide a name for the new key pair. Download and save the private key file (in the PEM format) in a safe location.

8. Click on the **Create Cluster** button to deploy the EMR HBase cluster.

Amazon will charge you for the compute and storage resources you use by clicking on **Create Cluster** in the preceding step. Refer to the *Saving money using Amazon EC2 Spot Instances to execute EMR job flows* recipe that we discussed earlier to find out how you can save money by using Amazon EC2 Spot Instances.

The following steps will show you how to connect to the master node of the deployed HBase cluster to start the HBase shell:

1. Go to the Amazon EMR console (`https://console.aws.amazon.com/elasticmapreduce`). Select the **Cluster details** for the HBase cluster to view more information about the cluster. Retrieve **Master Public DNS Name** from the information pane.

2. Use the master public DNS name and the EC2 PEM-based key (selected in step 4) to connect to the master node of the HBase cluster:

```
$ ssh -i ec2.pem hadoop@ec2-184-72-138-2.compute-
1.amazonaws.com
```

3. Start the HBase shell using the `hbase shell` command. Create a table named `'test'` in your HBase installation and insert a sample entry to the table using the `put` command. Use the `scan` command to view the contents of the table.

```
$ hbase shell

. . . . . . . . .


hbase(main):001:0> create 'test','cf'

0 row(s) in 2.5800 seconds


hbase(main):002:0> put 'test','row1','cf:a','value1'

0 row(s) in 0.1570 seconds


hbase(main):003:0> scan 'test'
ROW                      COLUMN+CELL
 row1                    column=cf:a, timestamp=1347261400477,
value=value1

1 row(s) in 0.0440 seconds


hbase(main):004:0> quit
```

The following step will back up the data stored in an Amazon EMR HBase cluster.

4. Execute the following command using the AWS CLI to schedule a periodic backup of the data stored in an EMR HBase cluster. Retrieve the cluster ID (for example, `j-FDMXCBZP9P85`) from the EMR console. Replace the `<cluster_id>` using the retrieved job flow name. Change the backup directory path (`s3://hcb-c2-data/hbase-backup`) according to your backup data Blob. Wait for several minutes for the backup to be performed.

```
$ aws emr schedule-hbase-backup --cluster-id <cluster_id> \
  --type full -dir s3://hcb-c2-data/hbase-backup \

--interval 1 --unit hours
```

5. Go to the **Cluster Details** page in the EMR console and click on **Terminate**.

   Now, we will start a new Amazon EMR HBase cluster by restoring data from a backup:

6. Create a new job flow by clicking on the **Create Cluster** button in the EMR console. Provide a name for your cluster. Provide a path in **Log folder S3 location** and select an AMI version with Hadoop v2 (for example, AMI version 3.1.0 with Hadoop 2.4.0).

7. Select **HBase** from the **Additional Applications** drop-down box under the **Applications to be installed** section. Click on **Configure and add**.

8. Configure the EMR HBase cluster to restore data from the previous data backup. Select the **Restore from Backup** option and provide the backup directory path you used in step 9 in the **Backup Location** textbox. You can leave the backup version textbox empty and the EMR would restore the latest backup. Click on **Continue**.

9. Repeat steps 4, 5, 6, and 7.

10. Start the HBase shell by logging in to? the master node of the new HBase cluster. Use the `list` command to list the set tables in HBase and the `scan 'test'` command to view the contents of the `'test'` table.

```
$ hbase shell
. . . . . . . . .


hbase(main):001:0> list
TABLE
test
1 row(s) in 1.4870 seconds


hbase(main):002:0> scan 'test'
ROW                      COLUMN+CELL
 row1                       column=cf:a, timestamp=1347318118294,
value=value1
1 row(s) in 0.2030 seconds
```

11. Terminate your cluster using the EMR console by going to the **Cluster Details** page and clicking on the **Terminate** button.

## See also

The HBase-related recipes in *Chapter 7, Hadoop Ecosystem II – Pig, HBase, Mahout, and Sqoop.*

# Using EMR bootstrap actions to configure VMs for the Amazon EMR jobs

**EMR bootstrap actions** provide us a mechanism to configure the EC2 instances before running our MapReduce computations. Examples of bootstrap actions include providing custom configurations for Hadoop, installing any dependent software, distributing a common dataset, and so on. Amazon provides a set of predefined bootstrap actions as well as allowing us to write our own custom bootstrap actions. EMR runs the bootstrap actions in each instance before Hadoop cluster services are started.

In this recipe, we are going to use a stop words list to filter out the common words from our WordCount sample. We download the stop words list to the workers using a custom bootstrap action.

## How to do it...

The following steps show you how to download a file to all the EC2 instances of an EMR computation using a bootstrap script:

1.  Save the following script to a file named `download-stopwords.sh`. Upload the file to a Blob container in the Amazon S3. This custom bootstrap file downloads a stop words list to each instance and copies it to a pre-designated directory inside the instance.

    ```
    #!/bin/bash

    set -e

    wget http://www.textfixer.com/resources/common-english-words-
    with-contractions.txt

    mkdir -p /home/hadoop/stopwords

    mv common-english-words-with-contractions.txt
    /home/hadoop/stopwords
    ```

2.  Complete steps 1 to 10 of the *Running Hadoop MapReduce computations using Amazon Elastic MapReduce* recipe in this chapter.

3.  Select the **Add Bootstrap Actions** option in the **Bootstrap Actions** tab. Select **Custom Action** in the **Add Bootstrap Actions** drop-down box. Click on **Configure and add**. Give a name to your action in the **Name** textbox and provide the S3 path of the location where you uploaded the `download-stopwords.sh` file in the **S3 location** textbox. Click on **Add**.

4.  Add **Steps** if needed.

5.  Click on the **Create Cluster** button to launch instances and to deploy the MapReduce cluster.

6.  Click on **Refresh** in the EMR console and go to your **Cluster Details** page to view the details of the cluster.

## There's more...

Amazon provides us with the following predefined bootstrap actions:

-   `configure-daemons`: This allows us to set **Java Virtual Machine** (**JVM**) options for the Hadoop daemons, such as the heap size and garbage collection behavior.

-   `configure-hadoop`: This allows us to modify the Hadoop configuration settings. Either we can upload a Hadoop configuration XML or we can specify individual configuration options as key-value pairs.

-   `memory-intensive`: This allows us to configure the Hadoop cluster for memory-intensive workloads.

-   `run-if`: This allows us to run bootstrap actions based on a property of an instance. This action can be used in scenarios where we want to run a command only in the Hadoop master node.

You can also create shutdown actions by writing scripts to a designated directory in the instance. Shutdown actions are executed after the job flow is terminated.

Refer to http://docs.amazonwebservices.com/ElasticMapReduce/latest/ DeveloperGuide/Bootstrap.html for more information.

# Using Apache Whirr to deploy an Apache Hadoop cluster in a cloud environment

Apache Whirr provides a set of cloud-vendor-neutral set of libraries to provision services on the cloud resources. Apache Whirr supports the provisioning, installing, and configuring of Hadoop clusters in several cloud environments. In addition to Hadoop, Apache Whirr also supports the provisioning of Apache Cassandra, Apache ZooKeeper, Apache HBase, Voldemort (key-value storage), and Apache Hama clusters on the cloud environments.

 The installation programs of several commercial Hadoop distributions, such as Hortonworks HDP and Cloudera CDH, now support installation and configuration of those distributions on Amazon EC2 instances. These commercial-distribution-based installations would provide you with a more feature-rich Hadoop cluster on the cloud than using Apache Whirr.

In this recipe, we are going to start a Hadoop cluster on Amazon EC2 using Apache Whirr and run the WordCount MapReduce sample (the *Writing a WordCount MapReduce application, bundling it, and running it using the Hadoop local mode* recipe from *Chapter 1, Getting Started with Hadoop v2*) program on that cluster.

## How to do it...

The following are the steps to deploy a Hadoop cluster on Amazon EC2 using Apache Whirr and to execute the WordCount MapReduce sample on the deployed cluster:

1. Download and unzip the Apache Whirr binary distribution from http://whirr. apache.org/. You may be able to install Whirr through your Hadoop distribution as well.

2. Run the following command from the extracted directory to verify your Whirr installation:

```
$ whirr version
Apache Whirr 0.8.2
jclouds 1.5.8
```

3. Export your AWS access keys to the `AWS_ACCESS_KEY_ID` and `AWS_SECRET_ACCESS_KEY` environment parameters:

```
$ export AWS_ACCESS_KEY_ID=AKIA…
$ export AWS_SECRET_ACCESS_KEY=…
```

4. Generate an **rsa key pair** using the following command. This key pair is not the same as your AWS key pair.

```
$ssh-keygen -t rsa -P ''
```

5. Locate the file named `recipes/hadoop-yarn-ec2.properties` in your Apache Whirr installation. Copy it to your working directory. Change the `whirr.hadoop.version` property to match a current Hadoop version (for example, 2.5.2) available in the Apache Hadoop downloads page.

6. If you provided a custom name for your key-pair in the previous step, change the `whirr.private-key-file` and the `whirr.public-key-file` property values in the `hadoop-yarn-ec2.properties` file to the paths of the private key and the public key you generated.

 The `whirr.aws-ec2-spot-price` property is an optional property that allows us to use cheaper EC2 Spot Instances. You can delete that property to use EC2 traditional on-demand instances.

7. Execute the following command pointing to your `hadoop-yarn-ec2.properties` file to launch your Hadoop cluster on EC2. After the successful cluster creation, this command outputs an SSH command that we can use to log in to the EC2 Hadoop cluster.

```
$bin/whirr launch-cluster --config hadoop-yarn-ec2.properties
```

8. The traffic from the outside to the provisioned EC2 Hadoop cluster is routed through the master node. Whirr generates a script that we can use to start this proxy, under a subdirectory named after your Hadoop cluster inside the `~/.whirr` directory. Run this in a new terminal. It will take a few minutes for Whirr to start the cluster and to generate this script.

```
$cd ~/.whirr/Hadoop-yarn/
$hadoop-proxy.sh
```

9. You can open the Hadoop web-based monitoring console in your local machine by configuring this proxy in your web browser.

10. Whirr generates a `hadoop-site.xml` file for your cluster in the `~/.whirr/<your cluster name>` directory. You can use it to issue Hadoop commands from your local machine to your Hadoop cluster on EC2. Export the path of the generated `hadoop-site.xml` to an environmental variable named `HADOOP_CONF_DIR`. Copy the `hadoop-site.xml` file in this directory to another file named `core-site.xml`. To execute the Hadoop commands, you should have Hadoop v2 binaries installed in your machine.

    ```
    $ cp ~/.whirr/hadoop-yarn/hadoop-site.xml ~/.whirr/hadoop-yarn/core-site.xml

    $ export HADOOP_CONF_DIR=~/.whirr/hadoop-yarn/

    $ hdfs dfs -ls /
    ```

11. Create a directory named `wc-input-data` in HDFS and upload a text dataset to that directory. Depending on the version of Whirr, you may have to create your home directory first.

    ```
    $ hdfs dfs -mkdir /user/<user_name>

    $ hdfs  dfs -mkdir wc-input-data

    $ hdfs dfs -put sample.txt wc-input-data
    ```

12. In this step, we run the Hadoop WordCount sample in the Hadoop cluster we started in Amazon EC2:

    ```
    $ hadoop jar hcb-c1-samples.jar chapter1.WordCount \
    wc-input-data wc-out
    ```

13. View the results of the WordCount computation by executing the following commands:

    ```
    $hadoop fs -ls wc-out
    Found 3 items
    -rw-r--r--    3 thilina supergroup          0 2012-09-05 15:40
    /user/thilina/wc-out/_SUCCESS
    drwxrwxrwx    - thilina supergroup          0 2012-09-05 15:39
    /user/thilina/wc-out/_logs
    -rw-r--r--    3 thilina supergroup      19908 2012-09-05 15:40
    /user/thilina/wc-out/part-r-00000


    $ hadoop fs -cat wc-out/part-* | more
    ```

14. Issue the following command to shut down the Hadoop cluster. Make sure to download any important data before shutting down the cluster, as the data will be permanently lost after shutting down the cluster.

    ```
    $bin/whirr destroy-cluster --config hadoop.properties
    ```

## How it works...

The following are the descriptions of the properties we used in the `hadoop.properties` file.

```
whirr.cluster-name=Hadoop-yarn
```

The preceding property provides a name for the cluster. The instances of the cluster will be tagged using this name.

```
whirr.instance-templates=1 hadoop-namenode+yarn-resource-
manager+mapreduce-historyserver, 1 hadoop-datanode+yarn-
nodemanager
```

This property specifies the number of instances to be used for each set of roles and the type of roles for the instances.

```
whirr.provider=aws-ec2
```

We use the Whirr Amazon EC2 provider to provision our cluster.

```
whirr.private-key-file=${sys:user.home}/.ssh/id_rsa
whirr.public-key-file=${sys:user.home}/.ssh/id_rsa.pub
```

Both the properties mentioned earlier point to the paths of the private key and the public key you provide for the cluster.

```
whirr.hadoop.version=2.5.2
```

We specify a custom Hadoop version using the preceding property.

```
whirr.aws-ec2-spot-price=0.15
```

This property specifies a bid price for the Amazon EC2 Spot Instances. Specifying this property triggers Whirr to use EC2 Spot Instances for the cluster. If the bid price is not met, Apache Whirr Spot Instance requests a time out after 20 minutes. Refer to the *Saving money using Amazon EC2 Spot Instances to execute EMR job flows* recipe for more details.

More details on Whirr configuration can be found at `http://whirr.apache.org/docs/0.8.1/configuration-guide.html`.

## See also

The *Saving money using Amazon EC2 Spot Instances to execute EMR job flows* recipe.

# 3
# Hadoop Essentials – Configurations, Unit Tests, and Other APIs

In this chapter, we will cover:

- Optimizing Hadoop YARN and MapReduce configurations for cluster deployments
- Shared user Hadoop clusters – using Fair and Capacity schedulers
- Setting classpath precedence to user-provided JARs
- Speculative execution of straggling tasks
- Unit testing Hadoop MapReduce applications using MRUnit
- Integration testing Hadoop MapReduce applications using MiniYarnCluster
- Adding a new DataNode
- Decommissioning DataNodes
- Using multiple disks/volumes and limiting HDFS disk usage
- Setting the HDFS block size
- Setting the file replication factor
- Using the HDFS Java API

# Introduction

This chapter describes how to perform advanced administration steps in your Hadoop cluster, how to develop unit and integration tests for Hadoop MapReduce programs and how to use the Java API of HDFS. This chapter assumes that you have followed the first chapter and have installed Hadoop in a clustered or pseudo-distributed setup.

**Sample code and data**

The sample code files for this book are available in GitHub at `https://github.com/thilg/hcb-v2`. The `chapter3` folder of the code repository contains the sample source code files for this chapter.

Sample codes can be compiled and built by issuing the `gradle build` command in the `chapter3` folder of the code repository. Project files for Eclipse IDE can be generated by running the `gradle eclipse` command in the main folder of the code repository. Project files for the IntelliJ IDEA IDE can be generated by running the `gradle idea` command in the main folder of the code repository.

# Optimizing Hadoop YARN and MapReduce configurations for cluster deployments

In this recipe, we explore some of the important configuration options of Hadoop YARN and Hadoop MapReduce. Commercial Hadoop distributions typically provide a GUI-based approach to specify Hadoop configurations.

YARN allocates resource containers to the applications based on the resource requests made by the applications and the available resource capacity of the cluster. A resource request by an application would consist of the number of containers required and the resource requirement of each container. Currently, most container resource requirements are specified using the amount of memory. Hence, our focus in this recipe will be mainly on configuring the memory allocation of a YARN cluster.

## Getting ready

Set up a Hadoop cluster by following the recipes in the first chapter.

## How to do it...

The following instructions will show you how to configure the memory allocation in a YARN cluster. The number of tasks per node is derived using this configuration:

1. The following property specifies the amount of memory (RAM) that can be used by YARN containers in a worker node. It's advisable to set this slightly less than the amount of physical RAM present in the node, leaving some memory for the OS and other non-Hadoop processes. Add or modify the following lines in the `yarn-site.xml` file:

   ```
   <property>
     <name>yarn.nodemanager.resource.memory-mb</name>
     <value>100240</value>
   </property>
   ```

2. The following property specifies the minimum amount of memory (RAM) that can be allocated to a YARN container in a worker node. Add or modify the following lines in the `yarn-site.xml` file to configure this property.

   If we assume that all the YARN resource-requests request containers with only the minimum amount of memory, the maximum number of concurrent resource containers that can be executed in a node equals *(YARN memory per node specified in step 1)/(YARN minimum allocation configured below)*. Based on this relationship, we can use the value of the following property to achieve the desired number of resource containers per node.

   The number of resource containers per node is recommended to be less than or equal to the minimum of *(2\*number CPU cores)* or *(2\* number of disks)*.

   ```
   <property>
     <name>yarn.scheduler.minimum-allocation-mb</name>
     <value>3072</value>
   </property>
   ```

3. Restart the YARN ResourceManager and NodeManager services by running `sbin/stop-yarn.sh` and `sbin/start-yarn.sh` from the `HADOOP_HOME` directory.

The following instructions will show you how to configure the memory requirements of the MapReduce applications.

1. The following properties define the maximum amount of memory (RAM) that will be available to each Map and Reduce task. These memory values will be used when MapReduce applications request resources from YARN for Map and Reduce task containers. Add the following lines to the `mapred-site.xml` file:

   ```
   <property>
     <name>mapreduce.map.memory.mb</name>
     <value>3072</value>
   </property>
   ```

```
<property>
  <name>mapreduce.reduce.memory.mb</name>
  <value>6144</value>
</property>
```

2. The following properties define the JVM heap size of the Map and Reduce tasks respectively. Set these values to be slightly less than the corresponding values in step **4**, so that they won't exceed the resource limits of the YARN containers. Add the following lines to the `mapred-site.xml` file:

```
<property>
  <name>mapreduce.map.java.opts</name>
  <value>-Xmx2560m</value>
</property>
<property>
  <name>mapreduce.reduce.java.opts</name>
  <value>-Xmx5120m</value>
</property>
```

## How it works...

We can control Hadoop configurations through the following four configuration files. Hadoop reloads the configurations from these configuration files after a cluster restart:

- `core-site.xml`: Contains the configurations common to the whole Hadoop distribution
- `hdfs-site.xml`: Contains configurations for HDFS
- `mapred-site.xml`: Contains configurations for MapReduce
- `yarn-site.xml`: Contains configurations for the YARN ResourceManager and NodeManager processes

Each configuration file has name-value pairs expressed in XML format, defining the configurations of different aspects of Hadoop. The following is an example of a property in a configuration file. The `<configuration>` tag is the top-level parent XML container and `<property>` tags, which define individual properties, are specified as child tags inside the `<configuration>` tag:

```
<configuration>
  <property>
    <name>mapreduce.reduce.shuffle.parallelcopies</name>
```

```
        <value>20</value>
    </property>
    ...
    </configuration>
```

Some configurations can be configured on a per-job basis using the `job.`
`getConfiguration().set(name, value)` method from the Hadoop MapReduce job
driver code.

## There's more...

There are many similar important configuration properties defined in Hadoop. The following
are some of them:

| conf/core-site.xml | | |
|---|---|---|
| **Name** | **Default value** | **Description** |
| `fs.inmemory.size.mb` | 200 | Amount of memory allocated to the in-memory filesystem that is used to merge map outputs at reducers in MBs |
| `io.file.buffer.size` | 131072 | Size of the read/write buffer used by sequence files |

| conf/mapred-site.xml | | |
|---|---|---|
| **Name** | **Default value** | **Description** |
| `mapreduce.reduce.shuffle.parallelcopies` | 20 | Maximum number of parallel copies the reduce step will execute to fetch output from many parallel jobs |
| `mapreduce.task.io.sort.factor` | 50 | Maximum number of streams merged while sorting files |
| `mapreduce.task.io.sort.mb` | 200 | Memory limit while sorting data in MBs |

| conf/hdfs-site.xml | | |
|---|---|---|
| **Name** | **Default value** | **Description** |
| `dfs.blocksize` | 134217728 | HDFS block size |
| `dfs.namenode.handler.count` | 200 | Number of server threads to handle RPC calls in NameNodes |

You can find a list of deprecated properties in the latest version of Hadoop and the new replacement properties for them at `http://hadoop.apache.org/docs/current/hadoop-project-dist/hadoop-common/DeprecatedProperties.html`.

The following documents provide the list of properties, their default values, and the descriptions of each of the configuration files mentioned earlier:

- **Common configuration**: `http://hadoop.apache.org/docs/current/hadoop-project-dist/hadoop-common/core-default.xml`

- **HDFS configuration**: `https://hadoop.apache.org/docs/current/hadoop-project-dist/hadoop-hdfs/hdfs-default.xml`

- **YARN configuration**: `http://hadoop.apache.org/docs/current/hadoop-yarn/hadoop-yarn-common/yarn-default.xml`

- **MapReduce configuration**: `http://hadoop.apache.org/docs/stable/hadoop-mapreduce-client/hadoop-mapreduce-client-core/mapred-default.xml`

# Shared user Hadoop clusters – using Fair and Capacity schedulers

The Hadoop YARN scheduler is responsible for assigning resources to the applications submitted by users. In Hadoop YARN, these can be any YARN application in addition to MapReduce applications. Currently, the default YARN resource allocation is based on the memory requirements of the application, while resource allocation based on other resources such as CPU can be configured additionally.

Hadoop YARN supports a pluggable scheduling framework, where the cluster administrator has the choice of selecting an appropriate scheduler for the cluster. By default, YARN supports a **First in First out** (**FIFO**) scheduler, which executes jobs in the same order as they arrive using a queue of jobs. However, FIFO scheduling might not be the best option for large multi-user Hadoop deployments, where the cluster resources have to be shared across different users and different applications to ensure maximum productivity from the cluster. Please note that commercial Hadoop distributions may use a different scheduler such as Fair scheduler (for example, Cloudera CDH) or Capacity scheduler (for example, Hortonworks HDP) as the default YARN scheduler.

In addition to the default FIFO scheduler, YARN contains the following two schedulers (if required, it is possible for you to write your own scheduler as well):

▸ **Fair scheduler**: The Fair scheduler allows all jobs to receive an equal share of resources. The resources are assigned to newly submitted jobs as and when the resources become available until all submitted and running jobs have the same amount of resources. The Fair scheduler ensures that short jobs are completed at a realistic speed, while not starving long-running larger jobs for longer periods. With the Fair scheduler, it's also possible to define multiple queues and queue hierarchies with guaranteed minimum resources to each queue, where the jobs in a particular queue share the resources equally. Resources allocated to any empty queues get divided among the queues with active jobs. The Fair scheduler also allows us to set job priorities, which are used to calculate the proportion of resource distribution inside a queue.

▸ **Capacity scheduler**: The Capacity scheduler allows a large cluster to be shared across multiple organizational entities while ensuring guaranteed capacity for each entity and that no single user or job holds all the resources. This allows organizations to achieve economies of scale by maintaining a centralized Hadoop cluster shared between various entities. In order to achieve this, the Capacity scheduler defines queues and queue hierarchies, with each queue having a guaranteed capacity. The Capacity scheduler allows the jobs to use the excess resources (if any) from the other queues.

## How to do it...

This recipe describes how to change the scheduler in Hadoop:

1. Shut down the Hadoop cluster.
2. Add the following to the `yarn-site.xml file`:

```
<property>
   <name>yarn.resourcemanager.scheduler.class</name>
   <value>org.apache.hadoop.yarn.server.resourcemanager.
scheduler.fair.FairScheduler</value>
</property>
```

3. Restart the Hadoop cluster.
4. Verify that the new scheduler has been applied by going to `http://<master-noe>:8088/cluster/scheduler` in your installation.

## How it works...

When you follow the aforementioned steps, Hadoop will load the new scheduler settings when it is started. The Fair scheduler shares an equal amount of resources between users unless it has been configured otherwise.

We can provide an XML formatted allocation file, defining the queues for the Fair scheduler, using the `yarn.scheduler.fair.allocation.file` property in the `yarn-site.xml` file.

More details about the Fair scheduler and its configurations can be found at `https://hadoop.apache.org/docs/current/hadoop-yarn/hadoop-yarn-site/FairScheduler.html`.

## There's more...

You can enable the Capacity scheduler by adding the following to the `yarn-site.xml file` and restarting the cluster:

```
<property>
    <name>yarn.resourcemanager.scheduler.class</name>
<value>org.apache.hadoop.yarn.server.resourcemanager.scheduler.
capacity.CapacityScheduler</value>
</property>
```

The Capacity scheduler can be configured using the `capacity-scheduler.xml` file in the Hadoop configuration directory of the ResourceManager node. Issue the following command in the YARN ResourceManager node to load the configuration and to refresh the queues:

**$ yarn rmadmin -refreshQueues**

More details about the Capacity scheduler and its configurations can be found at `http://hadoop.apache.org/docs/current/hadoop-yarn/hadoop-yarn-site/CapacityScheduler.html`.

# Setting classpath precedence to user-provided JARs

While developing Hadoop MapReduce applications, you may encounter scenarios where your MapReduce application requires a newer version of an auxiliary library that is already included in Hadoop. By default, Hadoop gives classpath precedence to the libraries included with Hadoop, which can result in conflicts with the version of the library you provide with your applications. This recipe shows you how to configure Hadoop to give classpath precedence to user-provided libraries.

## How to do it...

The following steps show you how to add external libraries to the Hadoop task classpath and how to provide precedence to user-supplied JARs:

1. Set the following property in the driver program of your MapReduce computation:

   ```
   job.getConfiguration().set
   ("mapreduce.job.user.classpath.first","true");
   ```

2. Use the `-libjars` option in the Hadoop command to provide your libraries, as follows:

```
$hadoop jar hcb-c3-samples.jar \
chapter3.WordCountWithTools \
-libjars guava-15.0.jar \
InDir OutDir …
```

## How it works...

Hadoop will copy the JARs specified by `-libjars` in to the Hadoop `DistributedCache` and they will be made available to the classpath of all the tasks belonging to this particular job. When `mapreduce.user.classpath.first` is set, the user-supplied JARs will be appended to the classpath before the default Hadoop JARs and Hadoop dependencies.

# Speculative execution of straggling tasks

One of the main advantages of using Hadoop MapReduce is the framework-managed fault tolerance. When performing a large-scale distributed computation, parts of the computation can fail due to external causes such as network failures, disk failures, and node failures. When Hadoop detects an unresponsive task or a failed task, Hadoop will re-execute those tasks in a new node.

A Hadoop cluster may consist of heterogeneous nodes, and as a result there can be very slow nodes as well as fast nodes. Potentially, a few slow nodes and the tasks executing on those nodes can dominate the execution time of a computation. Hadoop introduces speculative execution optimization to avoid these slow-running tasks, which are called **stragglers**.

When most of the Map (or Reduce) tasks of a computation are completed, the Hadoop speculative execution feature will schedule duplicate executions of the remaining slow tasks in available alternate nodes. The slowness of a task is decided relative to the running time taken by the other tasks of the same computation. From a set of duplicate tasks, Hadoop will choose the results from the first completed task and will kill any other duplicate executions of that task.

## How to do it...

By default, speculative executions are enabled in Hadoop for both Map and Reduce tasks. If such duplicate executions are undesirable for your computations for some reason, you can disable (or enable) speculative executions as follows:

1. Run the WordCount sample passing the following option as an argument:

```
$ hadoop jar hcb-c32-samples.jar chapter3.WordCountWithTools \
    -Dmapreduce.map.speculative=false \
    -Dmapreduce.reduce.speculative=false \
    /data/input1 /data/output1
```

2. However, the preceding command only works if the job implements the `org.apache.hadoop.util.Tools` interface. Otherwise, set these properties in the MapReduce driver program using the following methods:

   ❑ For the whole job, use `job.setSpeculativeExecution(boolean specExec)`

   ❑ For Map tasks, use `job.setMapSpeculativeExecution(boolean specExec)`

   ❑ For Reduce tasks, use `Job.setReduceSpeculativeExecution(boolean specExec)`

## There's more...

You can configure the maximum number of retry attempts for a task using the properties, `mapreduce.map.maxattempts` and `mapreduce.reduce.maxattempts`, for Map and Reduce tasks, respectively. Hadoop declares a task as a failure after it exceeds the given number of retries. You can also use the `JobConf.setMaxMapAttempts()` and `JobConf.setMaxReduceAttempts()` functions to configure these properties. The default value for these properties is 4.

# Unit testing Hadoop MapReduce applications using MRUnit

**MRUnit** is a JUnit-based Java library that allows us to unit test Hadoop MapReduce programs. This makes it easy to develop as well as to maintain Hadoop MapReduce code bases. MRUnit supports testing `Mappers and Reducers` separately as well as testing MapReduce computations as a whole. In this recipe, we'll be exploring all three testing scenarios. The source code for the test program used in this recipe is available in the `chapter3\test\chapter3\WordCountWithToolsTest.java` file in the Git repository.

## Getting ready

We use Gradle as the build tool for our sample code base. If you haven't already done so, please install Gradle by following the instructions given in the introduction section of *Chapter 1, Getting Started with Hadoop v2*.

## How to do it...

The following steps show you how to perform unit testing of a Mapper using MRUnit:

1. In the `setUp` method of the test class, initialize an MRUnit `MapDriver` instance with the Mapper class you want to test. In this example, we are going to test the Mapper of the WordCount MapReduce application we discussed in earlier recipes:

```
public class WordCountWithToolsTest {

    MapDriver<Object, Text, Text, IntWritable> mapDriver;

    @Before
    public void setUp() {
        WordCountWithTools.TokenizerMapper mapper =
            new WordCountWithTools.TokenizerMapper();
        mapDriver = MapDriver.newMapDriver(mapper);
    }
    ......
}
```

2. Write a test function to test the Mapper logic. Provide the test input to the Mapper using the `MapDriver.withInput` method. Then, provide the expected result of the Mapper execution using the `MapDriver.withOutput` method. Now, invoke the test using the `MapDriver.runTest` method. The `MapDriver.withAll` and `MapDriver.withAllOutput` methods allow us to provide a list of test inputs and a list of expected outputs, rather than adding them individually.

```
@Test
    public void testWordCountMapper() throws IOException {
        IntWritable inKey = new IntWritable(0);
        mapDriver.withInput(inKey, new Text("Test Quick"));
        ....
        mapDriver.withOutput(new Text("Test"),new
        IntWritable(1));
        mapDriver.withOutput(new Text("Quick"),new
        IntWritable(1));
        ...
        mapDriver.runTest();
    }
```

The following step shows you how to perform unit testing of a Reducer using MRUnit.

3. Similar to step 1 and 2, initialize a `ReduceDriver` by providing the Reducer class under test and then configure the `ReduceDriver` with the test input and the expected output. The input to the `reduce` function should conform to a key with a list of values. Also, in this test, we use the `ReduceDriver.withAllOutput` method to provide a list of expected outputs.

```
public class WordCountWithToolsTest {
  ReduceDriver<Text,IntWritable,Text,IntWritable>
  reduceDriver;

@Before
  public void setUp() {
    WordCountWithTools.IntSumReducer reducer =
      new WordCountWithTools.IntSumReducer();
    reduceDriver = ReduceDriver.newReduceDriver(reducer);
  }

@Test
  public void testWordCountReduce() throws IOException {
    ArrayList<IntWritable> reduceInList =
      new ArrayList<IntWritable>();
    reduceInList.add(new IntWritable(1));
    reduceInList.add(new IntWritable(2));

    reduceDriver.withInput(new Text("Quick"),
    reduceInList);
    ...
    ArrayList<Pair<Text, IntWritable>> reduceOutList =
      new ArrayList<Pair<Text,IntWritable>>();
    reduceOutList.add(new Pair<Text, IntWritable>
      (new Text("Quick"),new IntWritable(3)));
    ...
    reduceDriver.withAllOutput(reduceOutList);
    reduceDriver.runTest();
  }
}
```

The following steps show you how to perform unit testing on a whole MapReduce computation using MRUnit.

4. In this step, initialize a `MapReduceDriver` by providing the Mapper class and Reducer class of the MapReduce program that you want to test. Then, configure the `MapReduceDriver` with the test input data and the expected output data. When executed, this test will execute the MapReduce execution flow starting from the Map input stage to the Reduce output stage. It's possible to provide a combiner implementation to this test as well.

```java
public class WordCountWithToolsTest {
    ......
    MapReduceDriver<Object, Text, Text,
        IntWritable, Text,IntWritable> mapReduceDriver;

@Before
    public void setUp() {
        . . . .
        mapReduceDriver = MapReduceDriver.
            newMapReduceDriver(mapper, reducer);
    }

@Test
    public void testWordCountMapReduce() throws IOException {

        IntWritable inKey = new IntWritable(0);
        mapReduceDriver.withInput(inKey, new Text
            ("Test Quick"));
        ......
        ArrayList<Pair<Text, IntWritable>> reduceOutList
            = new ArrayList<Pair<Text,IntWritable>>();
        reduceOutList.add(new Pair<Text, IntWritable>
            (new Text("Quick"),new IntWritable(2)));
        ......
        mapReduceDriver.withAllOutput(reduceOutList);
        mapReduceDriver.runTest();
    }
}
```

5. The Gradle build script (or any other Java build mechanism) can be configured to execute these unit tests with every build. We can add the MRUnit dependency to the Gradle build (`chapter3/build.gradle`) file as follows:

```
dependencies {
    testCompile group: 'org.apache.mrunit', name: 'mrunit',
        version: '1.1.+',classifier: 'hadoop2'
    ......
}
```

6.  Use the following Gradle command to execute only the `WordCountWithToolsTest` unit test. This command executes any test class that matches the pattern `**/WordCountWith*.class`:

```
$ gradle -Dtest.single=WordCountWith test
:chapter3:compileJava UP-TO-DATE
:chapter3:processResources UP-TO-DATE
:chapter3:classes UP-TO-DATE
:chapter3:compileTestJava UP-TO-DATE
:chapter3:processTestResources UP-TO-DATE
:chapter3:testClasses UP-TO-DATE
:chapter3:test
BUILD SUCCESSFUL
Total time: 27.193 secs
```

7.  You can also execute MRUnit-based unit tests in your IDE. You can use the `gradle eclipse` or `gradle idea` commands to generate the project files for the Eclipse and IDEA IDE respectively.

## See also

▸  The *Integration testing Hadoop MapReduce applications using MiniYarnCluster* recipe in this chapter

▸  For more information about using MRUnit, go to `https://cwiki.apache.org/confluence/display/MRUNIT/MRUnit+Tutorial`

# Integration testing Hadoop MapReduce applications using MiniYarnCluster

While unit testing using MRUnit is very useful, there can be certain integration test scenarios that have to be tested in a cluster environment. MiniYARNCluster of Hadoop YARN is a cluster simulator that we can use to create a testing environment for such integration tests. In this recipe, we'll be using MiniYARNCluster to perform integration testing of the WordCountWithTools MapReduce application.

The source code for the test program used in this recipe is available in the `chapter3\test\chapter3\minicluster\WordCountMiniClusterTest.java` file in the Git repository.

## Getting ready

We use Gradle as the build tool for our sample code base. If you haven't already done so, please install Gradle by following the instructions given in the introduction section of *Chapter 1, Getting Started with Hadoop v2*. Export the JAVA_HOME environmental variable pointing to your JDK installation.

## How to do it...

The following steps show you how to perform integration testing of a MapReduce application using the MiniYarnCluster environment:

1. Within the setup method of your JUnit test, create an instance of MiniYarnCluster using MiniMRClientClusterFactory as follows. MiniMRClientCluster is a wrapper interface for MiniMRYarnCluster to provide support testing using Hadoop 1.x clusters.

   ```
   public class WordCountMiniClusterTest {
     private static MiniMRClientCluster mrCluster;
     private class InternalClass {
     }


   @BeforeClass
     public static void setup() throws IOException {
       // create the mini cluster to be used for the tests
       mrCluster = MiniMRClientClusterFactory.create
         (InternalClass.class, 1,new Configuration());
     }
   }
   ```

2. Make sure to stop the cluster inside the setup method of your test:

   ```
   @AfterClass
     public static void cleanup() throws IOException {
       // stopping the mini cluster
       mrCluster.stop();
     }
   ```

3. Within your test method, prepare a MapReduce computation using the configuration object of the `MiniYARNCluster` we just created. Submit the job and wait for its completion. Then test whether the job was successful.

```
@Test
  public void testWordCountIntegration() throws Exception{
......

    Job job = (new WordCountWithTools()).prepareJob
      (testInput,outDirString, mrCluster.getConfig());
    // Make sure the job completes successfully
    assertTrue(job.waitForCompletion(true));
    validateCounters
      (job.getCounters(), 12, 367, 201, 201);
  }
```

4. In this example, we will use the counters to validate the expected results of the MapReduce computation. You may also implement logic to compare the output data of the computation with the expected output of the computation. However, care must be taken to handle the possibility of having multiple output files due to the presence of multiple Reduce tasks.

```
private void validateCounters
  (Counters counters, long mapInputRecords,…) {

assertEquals("MapInputRecords",
  mapInputRecords, counters.findCounter(
  "org.apache.hadoop.mapred.Task$Counter",
  "MAP_INPUT_RECORDS").getValue());

    .........

}
```

5. Use the following Gradle command to execute only the `WordCountMiniClusterTest` JUnit test. This command executes any test class that matches the pattern `**/WordCountMini*.class`.

```
$ gradle -Dtest.single=WordCountMini test
:chapter3:compileJava UP-TO-DATE
:chapter3:processResources UP-TO-DATE
:chapter3:classes UP-TO-DATE
:chapter3:compileTestJava UP-TO-DATE
:chapter3:processTestResources UP-TO-DATE
:chapter3:testClasses UP-TO-DATE
:chapter3:test UP-TO-DATE

BUILD SUCCESSFUL
```

6. You can also execute `MiniYarnCluster`-based unit tests in your IDE. You can use the `gradle eclipse` or `gradle idea` commands to generate the project files for the Eclipse and IDEA IDE respectively.

## See also

▶ The *Unit testing Hadoop MapReduce applications using MRUnit* recipe in this chapter

▶ The *Hadoop counters for reporting custom metrics* recipe in *Chapter 4, Developing Complex Hadoop MapReduce Applications*

# Adding a new DataNode

This recipe shows you how to add new nodes to an existing HDFS cluster without restarting the whole cluster, and how to force HDFS to rebalance after the addition of new nodes. Commercial Hadoop distributions typically provide a GUI-based approach to add and remove DataNodes.

## Getting ready

1. Install Hadoop on the new node and replicate the configuration files of your existing Hadoop cluster. You can use `rsync` to copy the Hadoop configuration from another node; for example:

```
$ rsync -a <master_node_ip>:$HADOOP_HOME/etc/hadoop/
$HADOOP_HOME/etc/hadoop
```

2. Ensure that the master node of your Hadoop/HDFS cluster can perform password-less SSH to the new node. Password-less SSH setup is optional if you are not planning to use the `bin/*.sh` scripts from the master node to start/stop the cluster.

## How to do it...

The following steps will show you how to add a new DataNode to an existing HDFS cluster:

1. Add the IP or the DNS of the new node to the `$HADOOP_HOME/etc/hadoop/slaves` file in the master node.

2. Start the DataNode on the newly added slave node by using the following command:

```
$ $HADOOP_HOME/sbin/hadoop-deamons.sh start datanode
```

> You can also use the `$HADOOP_HOME/sbin/start-dfs.sh` script from the master node to start the DataNode daemons in the newly added nodes. This is helpful if you are adding more than one new DataNode to the cluster.

3. Check $HADOOP_HOME/logs/hadoop-*-datanode-*.log in the new slave node for any errors.

These steps apply to both adding a new node as well as rejoining a node that has crashed and restarted.

## There's more...

Similarly, you can add a new node to the Hadoop YARN cluster as well:

1. Start the NodeManager in the new node using the following command:

   ```
   > $HADOOP_HOME/sbin/yarn-deamons.sh start nodemanager
   ```

2. Check $HADOOP_HOME/logs/yarn-*-nodemanager-*.log in the new slave node for any errors.

## Rebalancing HDFS

When you add new nodes, HDFS will not rebalance automatically. However, HDFS provides a rebalancer tool that can be invoked manually. This tool will balance the data blocks across clusters up to an optional threshold percentage. Rebalancing would be very helpful if you are having space issues in the other existing nodes.

1. Execute the following command:

   ```
   > $HADOOP_HOME/sbin/start-balancer.sh –threshold 15
   ```

   The (optional) –threshold parameter specifies the percentage of disk capacity leeway to consider when identifying a node as under- or over-utilized. An under-utilized DataNode is a node whose utilization is less than *(average utilization-threshold)*. An over-utilized DataNode is a node whose utilization is greater than (average utilization + threshold). Smaller threshold values will achieve more evenly balanced nodes, but will take more time for the rebalancing. The default threshold value is 10 percent.

2. Rebalancing can be stopped by executing the sbin/stop-balancer.sh command.

3. A summary of the rebalancing is available in the $HADOOP_HOME/logs/hadoop-*-balancer*.out file.

## See also

The *Decommissioning DataNodes* recipe in this chapter.

# Decommissioning DataNodes

There can be multiple situations where you want to decommission one or more DataNodes from an HDFS cluster. This recipe shows how to gracefully decommission DataNodes without incurring data loss.

## How to do it...

The following steps show you how to decommission DataNodes gracefully:

1. If your cluster doesn't have it, add an `exclude` file to the cluster. Create an empty file in the NameNode and point to it from the `$HADOOP_HOME/etc/hadoop/hdfs-site.xml` file by adding the following property. Restart the NameNode:

   ```
   <property>
    <name>dfs.hosts.exclude</name>
    <value>FULL_PATH_TO_THE_EXCLUDE_FILE</value>
    <description>Names a file that contains a list of hosts
   that are not permitted to connect to the namenode. The
   full pathname of the file must be specified. If the value
   is empty, no hosts are excluded.</description>
   </property>
   ```

2. Add the hostnames of the nodes that are to be decommissioned to the `exclude` file.

3. Run the following command to reload the NameNode configuration:

   **$ hdfs dfsadmin -refreshNodes**

   This will start the decommissioning process. This process can take a significant amount of time as it requires replication of data blocks without overwhelming the other tasks of the cluster.

4. The progress of the decommissioning process is shown in the HDFS UI under the **Decommissioning Nodes** page. The progress can be monitored using the following command as well. Do not shut down the nodes until the decommissioning is complete.

   **$ hdfs dfsadmin -report**

   ```
   .....

   .....

   Name: myhost:50010

   Decommission Status : Decommission in progress

   Configured Capacity: ....

   .....
   ```

5. You can remove the nodes from the `exclude` file and execute the `hdfs dfsadmin -refreshNodes` command when you want to add the nodes back in to the cluster.

6. The decommissioning process can be stopped by removing the node name from the `exclude` file and then executing the `hdfs dfsadmin -refreshNodes` command.

## How it works...

When a node is in the decommissioning process, HDFS replicates the blocks in that node to the other nodes in the cluster. Decommissioning can be a slow process as HDFS purposely does it slowly to avoid overwhelming the cluster. Shutting down nodes without decommissioning may result in data loss.

After the decommissioning is complete, the nodes mentioned in the exclude file are not allowed to communicate with the NameNode.

## See also

The *Rebalancing HDFS* section of the *Adding a new DataNode* recipe in this chapter.

# Using multiple disks/volumes and limiting HDFS disk usage

Hadoop supports specifying multiple directories for the DataNode data directory. This feature allows us to utilize multiple disks/volumes to store data blocks in DataNodes. Hadoop tries to store equal amounts of data in each directory. It also supports limiting the amount of disk space used by HDFS.

## How to do it...

The following steps will show you how to add multiple disk volumes:

1. Create HDFS data storage directories in each volume.

2. Locate the `hdfs-site.xml` configuration file. Provide a comma-separated list of directories corresponding to the data storage locations in each volume under the `dfs.datanode.data.dir` property as follows:

```
<property>
        <name>dfs.datanode.data.dir</name>
        <value>/u1/hadoop/data, /u2/hadoop/data</value>
</property>
```

3. In order to limit disk usage, add the following property to the `hdfs-site.xml` file to reserve space for non-DFS usage. The value specifies the number of bytes that HDFS cannot use per volume:

```
<property>
   <name>dfs.datanode.du.reserved</name>
   <value>6000000000</value>
   <description>Reserved space in bytes per volume. Always
      leave this much space free for non dfs use.
   </description>
</property>
```

# Setting the HDFS block size

HDFS stores files across the cluster by breaking them down in to coarser-grained, fixed-size blocks. The default HDFS block size is 64 MB. Block size of a data product can affect the performance of the filesystem operations where larger block sizes would be more effective if you are storing and processing very large files. Block size of a data product can also affect the performance of MapReduce computations, as the default behavior of Hadoop is to create one Map task for each data block of the input files.

## How to do it...

The following steps show you how to use the NameNode configuration file to set the HDFS block size:

1. Add or modify the following code in the `$HADOOP_HOME/etc/hadoop/hdfs-site. xml` file. The block size is provided using the number of bytes. This change would not change the block size of the files that are already in the HDFS. Only the files copied after the change will have the new block size.

   NOT RETROACTIVE

```
<property>
        <name>dfs.blocksize</name>
        <value>134217728</value>
</property>
```

2. You can specify different HDFS block sizes for specific file paths as well. You can also specify the block size when uploading a file to HDFS from the command line as follows:

```
$ hdfs dfs \
 -Ddfs.blocksize=134217728 \
 -put data.in foo/test
```

## There's more...

You can also specify the block size when creating files using the HDFS Java API as well, in the following manner:

```
public FSDataOutputStream create(Path f,boolean overwrite, int
bufferSize, short replication,long blockSize)
```

You can use the `fsck` command to find the block size and block locations of a particular file path in the HDFS. You can find this information by browsing the filesystem from the HDFS monitoring console as well.

```
> $HADOOP_HOME/bin/hdfs fsck \
  /user/foo/data.in \
  -blocks -files -locations
......
/user/foo/data.in 215227246 bytes, 2 block(s): ....
0. blk_6981535920477261584_1059 len=134217728 repl=1 [hostname:50010]
1. blk_-8238102374790373371_1059 len=81009518 repl=1 [hostname:50010]


......
```

## See also

The *Setting the file replication factor* recipe in this chapter.

# Setting the file replication factor

HDFS stores files across the cluster by breaking them down into coarser-grained, fixed-size blocks. These coarser-grained data blocks are replicated to different DataNodes mainly for fault-tolerance purposes. Data block replication also has the ability to increase the data locality of the MapReduce computations and to increase the total data access bandwidth as well. Reducing the replication factor helps save storage space in HDFS.

**The HDFS replication factor** is a file-level property that can be set on a per-file basis. This recipe shows you how to change the default replication factor of an HDFS deployment affecting the new files that will be created afterwards, how to specify a custom replication factor at the time of file creation in HDFS, and how to change the replication factor of existing files in HDFS.

## How to do it...

Follow these instructions to set the file replication factor using the NameNode configuration:

1. Add or modify the `dfs.replication` property in `$HADOOP_HOME/etc/hadoop/hdfs-site.xml`. This change will not change the replication factor of the files that are already in the HDFS. Only the files copied after the change will have the new replication factor. Please be aware that reducing the replication factor decreases the reliability of the stored files and may also cause a performance decrease when processing that data as well.

   ```
   <property>
           <name>dfs.replication</name>
           <value>2</value>
   </property>
   ```

2. Set the file replication factor when uploading the files. You can specify the replication factor when uploading the file from the command line as follows:

   ```
   $ hdfs dfs \
    -Ddfs.replication=1 \
    -copyFromLocal \
    non-critical-file.txt /user/foo
   ```

3. Change the file replication factor of the existing file paths. The `setrep` command can be used to change the replication factor of files or file paths that are already in the HDFS in the following manner:

   ```
   $ hdfs dfs \
    -setrep 2 non-critical-file.txt


   Replication 2 set: hdfs://myhost:9000/user/foo/non-critical-file.txt
   ```

## How it works...

Have a look at the following command:

```
hdfs dfs -setrep [-R] <path>
```

The `<path>` parameter of the `setrep` command specifies the HDFS path where the replication factor has to be changed. The `-R` option recursively sets the replication factor for files and directories within a directory.

## There's more...

The replication factor of a file is displayed when listing the files using the `ls` command:

```
$ hdfs fs -ls
Found 1 item
-rw-r--r-- 2 foo supergroup ... /user/foo/non-critical-
file.txt
```

The replication factor of files is displayed in the HDFS monitoring the UI as well.

## See also

The *Setting the HDFS block size* recipe in this chapter.

# Using the HDFS Java API

**The HDFS Java API** can be used to interact with HDFS from any Java program. This API gives us the ability to utilize the data stored in HDFS from other Java programs as well as to process that data with other non-Hadoop computational frameworks. Occasionally, you may also come across a use case where you want to access HDFS directly from within a MapReduce application. However, if you are writing or modifying files in HDFS directly from a Map or Reduce task, please be aware that you are violating the side-effect-free nature of MapReduce, which might lead to data consistency issues based on your use case.

## How to do it...

The following steps show you how to use the HDFS Java API to perform filesystem operations on an HDFS installation using a Java program:

1. The following sample program creates a new file in HDFS, writes some text in the newly created file, and reads the file back from HDFS:

```java
import java.io.IOException;

import org.apache.hadoop.conf.Configuration;
import org.apache.hadoop.fs.FSDataInputStream;
import org.apache.hadoop.fs.FSDataOutputStream;
import org.apache.hadoop.fs.FileSystem;
import org.apache.hadoop.fs.Path;

public class HDFSJavaAPIDemo {
```

```
public static void main(String[] args) throws IOException
{
    Configuration conf = new Configuration();
    FileSystem fs = FileSystem.get(conf);
    System.out.println(fs.getUri());

    Path file = new Path("demo.txt");

    if (fs.exists(file)) {
        System.out.println("File exists.");
    } else {
        // Writing to file
        FSDataOutputStream outStream = fs.create(file);
        outStream.writeUTF("Welcome to HDFS Java API!!!");
        outStream.close();
    }

    // Reading from file
    FSDataInputStream inStream = fs.open(file);
    String data = inStream.readUTF();
    System.out.println(data);
    inStream.close();

    fs.close();
}
```

2.  Compile and package the preceding program by issuing the `gradle build` command in the `chapter3` folder of the source repository. The `hcb-c3-samples.jar` file will be created in the `build/libs` folder.

3.  You can execute the preceding sample using the following command. Running this sample using the `hadoop` script ensures that it uses the currently configured HDFS and the necessary dependencies from the Hadoop *classpath*.

```
$ hadoop jar \
  hcb-c3-samples.jar \
  chapter3.hdfs.javaapi.HDFSJavaAPIDemo


hdfs://yourhost:9000
Welcome to HDFS Java API!!!
```

4. Use the `ls` command to list the newly created file, shown as follows:

```
$ hdfs dfs -ls
Found 1 items
-rw-r--r--   3 foo supergroup         20 2012-04-27 16:57
/user/foo/demo.txt
```

## How it works...

In order to interact with HDFS programmatically, we first need to obtain a handle to the currently configured filesystem. For this, we instantiate a `Configuration` object and obtain a `FileSystem` handle, which will point to the HDFS NameNode of the Hadoop environment where we run this program. Several alternative methods to configure a `FileSystem` object have been discussed in the Configuring the FileSystem object section in this chapter:

```
Configuration conf = new Configuration();
FileSystem fs = FileSystem.get(conf);
```

The `FileSystem.create(filePath)` method creates a new file in the given path and provides us with an `FSDataOutputStream` object to the newly created file. `FSDataOutputStream` wraps `java.io.DataOutputStream` and allows the program to write primitive Java data types to the file. The `FileSystem.Create()` method overrides if the file exists. In this example, the file will be created relative to your HDFS home directory, which would result in a path similar to `/user/<user_name>/demo.txt`. Your HDFS home directory has to be created beforehand.

```
Path file = new Path("demo.txt");
FSDataOutputStream outStream = fs.create(file);
outStream.writeUTF("Welcome to HDFS Java API!!!");
outStream.close();
```

`FileSystem.open(filepath)` opens an `FSDataInputStream` to the given file. `FSDataInputStream` wraps `java.io.DataInputStream` and allows the program to read primitive Java data types from the file.

```
FSDataInputStream inStream = fs.open(file);
String data = inStream.readUTF();
System.out.println(data);
inStream.close();
```

## There's more...

The HDFS Java API supports many more filesystem operations than we have used in the preceding sample. The full API documentation can be found at `http://hadoop.apache.org/docs/current/api/org/apache/hadoop/fs/FileSystem.html`.

## Configuring the FileSystem object

We can use the HDFS Java API from outside the Hadoop environment as well. When doing so, we have to explicitly configure the HDFS NameNode and the port. The following are a couple of ways to perform that configuration:

▸ You can load the configuration files to the `configuration` object before retrieving the `FileSystem` object as follows. Make sure to add all the Hadoop and dependency libraries to the classpath.

```
Configuration conf = new Configuration();
conf.addResource(new Path("/etc/hadoop/core-site.xml"));
conf.addResource(new Path("/etc/hadoop/conf/hdfs-site.xml"));
FileSystem fileSystem = FileSystem.get(conf);
```

▸ You can also specify the NameNode and the port as follows. Replace the `NAMENODE_HOSTNAME` and `PORT` with the hostname and the port of the NameNode of your HDFS installation.

```
Configuration conf = new Configuration();
conf.set("fs.defaultFS, "hdfs://NAMENODE_HOSTNAME:PORT");
FileSystem fileSystem = FileSystem.get(conf);
```

The HDFS filesystem API is an abstraction that supports several filesystems. If the preceding program does not find a valid HDFS configuration, it will point to the local filesystem instead of the HDFS. You can identify the current filesystem of the `fileSystem` object using the `getUri()` function as follows. It would result in `hdfs://your_namenode:port` if it's using a properly configured HDFS and `file:///` if it is using the local filesystem.

```
fileSystem.getUri();
```

## Retrieving the list of data blocks of a file

The `getFileBlockLocations()` function of the `fileSystem` object allows you to retrieve the list of data blocks of a file stored in HDFS, together with hostnames where the blocks are stored and the block offsets. This information would be very useful if you are planning on doing any local operations on the file data using a framework other than Hadoop MapReduce.

```
FileStatus fileStatus = fs.getFileStatus(file);
BlockLocation[] blocks = fs.getFileBlockLocations(
    fileStatus, 0, fileStatus.getLen());
```

# 4

# Developing Complex Hadoop MapReduce Applications

In this chapter, we will cover the following recipes:

- ▸ Choosing appropriate Hadoop data types
- ▸ Implementing a custom Hadoop Writable data type
- ▸ Implementing a custom Hadoop key type
- ▸ Emitting data of different value types from a Mapper
- ▸ Choosing a suitable Hadoop InputFormat for your input data format
- ▸ Adding support for new input data formats – implementing a custom InputFormat
- ▸ Formatting the results of MapReduce computations – using Hadoop OutputFormats
- ▸ Writing multiple outputs from a MapReduce computation
- ▸ Hadoop intermediate data partitioning
- ▸ Secondary sorting – sorting Reduce input values
- ▸ Broadcasting and distributing shared resources to tasks in a MapReduce job – Hadoop DistributedCache
- ▸ Using Hadoop with legacy applications – Hadoop streaming
- ▸ Adding dependencies between MapReduce jobs
- ▸ Hadoop counters for reporting custom metrics

# Introduction

This chapter introduces you to several advanced Hadoop MapReduce features that will help you to develop highly customized, efficient MapReduce applications.

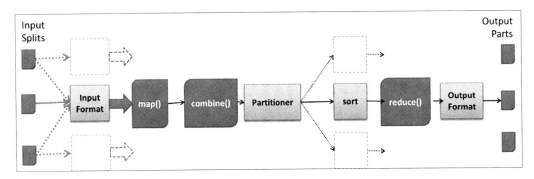

The preceding figure depicts the typical flow of a Hadoop MapReduce computation. The InputFormat reads the input data from HDFS and parses the data to create key-value pair inputs for the map function. InputFormat also performs the logical partitioning of data to create the Map tasks of the computation. A typical MapReduce computation creates a Map task for each input HDFS data block. Hadoop invokes the user provided map function for each of the generated key-value pairs. As mentioned in *Chapter 1, Getting Started with Hadoop v2*, if provided, the optional combiner step may get invoked with the output data from the map function.

The **Partitioner** step then partitions the output data of the Map task in order to send them to the respective Reduce tasks. This partitioning is performed using the key field of the Map task output key-value pairs and results in a number of partitions equal to the number of Reduce tasks. Each Reduce task fetches the respective output data partitions from the Map tasks (also known as **shuffling**) and performs a merge sort of the data based on the key field. Hadoop also groups the input data to the reduce function based on the key field of the data before invoking the reduce function. The output key-value pairs from the Reduce task would get written to the HDFS based on the format specified by the OutputFormat class.

In this chapter, we will explore the different parts of the earlier mentioned high-level flow of a Hadoop MapReduce computation in detail and explore the options and customizations available for each step. First you'll learn the different data types provided by Hadoop and the steps to implement custom data types for Hadoop MapReduce computations. Then we'll walk through the different data InputFormats and OutputFormats provided by Hadoop. Next, we will get a basic understanding of how to add support for new data formats in Hadoop as well as mechanisms for outputting more than one data product from a single MapReduce computation. We will also explore the Map output data partitioning and use that knowledge to introduce secondary sorting of the reduce function input data values.

In addition to the above, we will also discuss other advanced Hadoop features such as using **DistributedCache** for distributing the data, using Hadoop streaming feature for quick prototyping of Hadoop computations, and using Hadoop counters to report custom metrics for your computations as well as adding job dependencies to manage simple DAG-based workflows of Hadoop MapReduce computations.

**Sample code and data**

The example code files for this book are available in GitHub at `https://github.com/thilg/hcb-v2`. The `chapter4` folder of the code repository contains the sample source code files for this chapter.

You can download the data for the log processing sample from `http://ita.ee.lbl.gov/html/contrib/NASA-HTTP.html`. You can find a description of the structure of this data from `http://ita.ee.lbl.gov/html/contrib/NASA-HTTP.html`. A small extract of this dataset that can be used for testing is available in the code repository at `chapter4/resources`.

Sample codes can be compiled by issuing the `gradle build` command in the `chapter4` folder of the code repository. Project files for Eclipse IDE can be generated by running the `gradle eclipse` command in the main folder of the code repository. Project files for IntelliJ IDEA IDE can be generated by running the `gradle idea` command in the main folder of the code repository.

# Choosing appropriate Hadoop data types

Hadoop uses the Writable interface-based classes as the data types for the MapReduce computations. These data types are used throughout the MapReduce computational flow, starting with reading the input data, transferring intermediate data between Map and Reduce tasks, and finally, when writing the output data. Choosing the appropriate `Writable` data types for your input, intermediate, and output data can have a large effect on the performance and the programmability of your MapReduce programs.

In order to be used as a `value` data type of a MapReduce computation, a data type must implement the `org.apache.hadoop.io.Writable` interface. The `Writable` interface defines how Hadoop should serialize and de-serialize the values when transmitting and storing the data. In order to be used as a `key` data type of a MapReduce computation, a data type must implement the `org.apache.hadoop.io.WritableComparable<T>` interface. In addition to the functionality of the `Writable` interface, the `WritableComparable` interface further defines how to compare the key instances of this type with each other for sorting purposes.

**Hadoop's Writable versus Java's Serializable**

Hadoop's Writable-based serialization framework provides a more efficient and customized serialization and representation of the data for MapReduce programs than using the general-purpose Java's native serialization framework. As opposed to Java's serialization, Hadoop's Writable framework does not write the type name with each object expecting all the clients of the serialized data to be aware of the types used in the serialized data. Omitting the type names makes the serialization process faster and results in compact, random accessible serialized data formats that can be easily interpreted by non-Java clients. Hadoop's Writable-based serialization also has the ability to reduce the object-creation overhead by reusing the `Writable` objects, which is not possible with Java's native serialization framework.

## How to do it...

The following steps show you how to configure the input and output data types of your Hadoop MapReduce application:

1. Specify the data types for the input (key: `LongWritable`, value: `Text`) and output (key: `Text`, value: `IntWritable`) key-value pairs of your Mapper using the generic-type variables:

```
public class SampleMapper extends Mapper
  <LongWritable, Text, Text, IntWritable> {

  public void map(LongWritable key, Text value,
    Context context) … {
  ……  }
}
```

2. Specify the data types for the input (key: `Text`, value: `IntWritable`) and output (key: `Text`, value: `IntWritable`) key-value pairs of your Reducer using the generic-type variables. The Reducer's input key-value pair data types should match the Mapper's output key-value pairs.

```
public class Reduce extends Reducer<Text, IntWritable,
  Text, IntWritable> {

  public void reduce(Text key,
    Iterable<IntWritable> values, Context context) {
  ……  }
}
```

3. Specify the output data types of the MapReduce computation using the `Job` object as shown in the following code snippet. These data types will serve as the output types for both the Reducer and the Mapper, unless you specifically configure the Mapper output types as in step 4.

```
Job job = new Job(..);
….
job.setOutputKeyClass(Text.class);
job.setOutputValueClass(IntWritable.class);
```

4. Optionally, you can configure the different data types for the Mapper's output key-value pairs using the following steps, when your Mapper and Reducer have different data types for the output key-value pairs.

```
job.setMapOutputKeyClass(Text.class);
job.setMapOutputValueClass(IntWritable.class);
```

## There's more...

Hadoop provides several primitive data types such as `IntWritable`, `LongWritable`, `BooleanWritable`, `FloatWritable`, and `ByteWritable`, which are the `Writable` versions of their respective Java primitive data types. We can use these types as both the `key` types as well as the `value` types.

The following are several more Hadoop built-in data types that we can use as both the `key` as well as the `value` types:

- `Text`: This stores UTF8 text
- `BytesWritable`: This stores a sequence of bytes
- `VIntWritable` and `VLongWritable`: These store variable length integer and long values
- `NullWritable`: This is a zero-length Writable type that can be used when you don't want to use a `key` or `value` type

The following Hadoop built-in collection data types can only be used as `value` types:

- `ArrayWritable`: This stores an array of values belonging to a `Writable` type. To use `ArrayWritable` type as the `value` type of a Reducer's input, you need to create a subclass of `ArrayWritable` to specify the type of the `Writable` values stored in it.

```
public class LongArrayWritable extends ArrayWritable {
  public LongArrayWritable() {
    super(LongWritable.class);
  }
}
```

▶ `TwoDArrayWritable`: This stores a matrix of values belonging to the same `Writable` type. To use the `TwoDArrayWritable` type as the `value` type of a Reducer's input, you need to specify the type of the stored values by creating a subclass of the `TwoDArrayWritable` type similar to the `ArrayWritable` type.

```
public class LongTwoDArrayWritable extends
TwoDArrayWritable {
  public LongTwoDArrayWritable() {
    super(LongWritable.class);
  }
}
```

▶ `MapWritable`: This stores a map of key-value pairs. Keys and values should be of the `Writable` data types. You can use the `MapWritable` function as follows. However, you should be aware that the serialization of `MapWritable` adds a slight performance penalty due to the inclusion of the class names of each object stored in the map.

```
MapWritable valueMap = new MapWritable();
valueMap.put(new IntWritable(1),new Text("test"));
```

▶ `SortedMapWritable`: This stores a sorted map of key-value pairs. Keys should implement the `WritableComparable` interface. Usage of `SortedMapWritable` is similar to the `MapWritable` function.

## See also

▶ The *Implementing a custom Hadoop Writable data type* recipe
▶ The *Implementing a custom Hadoop key type* recipe

# Implementing a custom Hadoop Writable data type

There can be use cases where none of the inbuilt data types match your requirement or a custom data type optimized for your use case may perform better than a Hadoop built-in data type. In such scenarios, we can easily write a custom `Writable` data type by implementing the `org.apache.hadoop.io.Writable` interface to define the serialization format of your data type. The `Writable` interface-based types can be used as `value` types in Hadoop MapReduce computations.

In this recipe, we implement a sample Hadoop `Writable` data type for HTTP server log entries. For the purpose of this sample, we consider that a log entry consists of the five fields: request host, timestamp, request URL, response size, and the HTTP status code. The following is a sample log entry:

```
192.168.0.2 - - [01/Jul/1995:00:00:01 -0400] "GET /history/apollo/
HTTP/1.0" 200 6245
```

You can download a sample HTTP server log dataset from `ftp://ita.ee.lbl.gov/traces/NASA_access_log_Jul95.gz`.

## How to do it...

The following are the steps to implement a custom Hadoop `Writable` data type for the HTTP server log entries:

1. Write a new `LogWritable` class implementing the `org.apache.hadoop.io.Writable` interface:

```
public class LogWritable implements Writable{

    private Text userIP, timestamp, request;
    private IntWritable responseSize, status;

    public LogWritable() {
        this.userIP = new Text();
        this.timestamp=  new Text();
        this.request = new Text();
        this.responseSize = new IntWritable();
        this.status = new IntWritable();
    }
    public void readFields(DataInput in) throws IOException {
        userIP.readFields(in);
        timestamp.readFields(in);
        request.readFields(in);
        responseSize.readFields(in);
        status.readFields(in);
    }

    public void write(DataOutput out) throws IOException {
        userIP.write(out);
        timestamp.write(out);
        request.write(out);
```

```
        responseSize.write(out);
        status.write(out);
    }

    ......... // getters and setters for the fields
    }
```

2.  Use the new `LogWritable` type as a `value` type in your MapReduce computation. In the following example, we use the `LogWritable` type as the Map output value type:

```
public class LogProcessorMap extends Mapper<LongWritable,
Text, Text, LogWritable> {
....
}

public class LogProcessorReduce extends Reducer<Text,
LogWritable, Text, IntWritable> {

    public void reduce(Text key,
    Iterable<LogWritable> values, Context context) {
        ......   }
}
```

3.  Configure the output types of the job accordingly.

```
Job job = ......
....
job.setOutputKeyClass(Text.class);
job.setOutputValueClass(IntWritable.class);
job.setMapOutputKeyClass(Text.class);
job.setMapOutputValueClass(LogWritable.class);
```

## How it works...

The `Writable` interface consists of the two methods, `readFields()` and `write()`. Inside the `readFields()` method, we de-serialize the input data and populate the fields of the `Writable` object.

```
public void readFields(DataInput in) throws IOException {
    userIP.readFields(in);
    timestamp.readFields(in);
    request.readFields(in);
    responseSize.readFields(in);
    status.readFields(in);
}
```

In the preceding example, we use the `Writable` types as the fields of our custom `Writable` type and use the `readFields()` method of the fields for de-serializing the data from the `DataInput` object. It is also possible to use Java primitive data types as the fields of the `Writable` type and to use the corresponding read methods of the `DataInput` object to read the values from the underlying stream as done in the following code snippet:

```
int responseSize = in.readInt();
String userIP = in.readUTF();
```

Inside the `write()` method, we write the fields of the `Writable` object to the underlying stream.

```
public void write(DataOutput out) throws IOException {
    userIP.write(out);
    timestamp.write(out);
    request.write(out);
    responseSize.write(out);
    status.write(out);
}
```

In case you are using Java primitive data types as the fields of the `Writable` object, then you can use the corresponding write methods of the `DataOutput` object to write the values to the underlying stream as follows:

```
out.writeInt(responseSize);
out.writeUTF(userIP);
```

## There's more...

Please be cautious about the following issues when implementing your custom `Writable` data type:

- In case you are adding a custom constructor to your custom `Writable` class, make sure to retain the default empty constructor.

- `TextOutputFormat` uses the `toString()` method to serialize the `key` and `value` types. In case you are using the `TextOutputFormat` to serialize the instances of your custom `Writable` type, make sure to have a meaningful `toString()` implementation for your custom `Writable` data type.

- While reading the input data, Hadoop may reuse an instance of the `Writable` class repeatedly. You should not rely on the existing state of the object when populating it inside the `readFields()` method.

## See also

The *Implementing a custom Hadoop key type* recipe.

# Implementing a custom Hadoop key type

The instances of Hadoop MapReduce `key` types should have the ability to compare against each other for sorting purposes. In order to be used as a `key` type in a MapReduce computation, a Hadoop `Writable` data type should implement the `org.apache.hadoop.io.WritableComparable<T>` interface. The `WritableComparable` interface extends the `org.apache.hadoop.io.Writable` interface and adds the `compareTo()` method to perform the comparisons.

In this recipe, we modify the `LogWritable` data type of the *Implementing a custom Hadoop Writable data type* recipe to implement the `WritableComparable` interface.

## How to do it...

The following are the steps to implement a custom Hadoop `WritableComparable` data type for the HTTP server log entries, which uses the request hostname and timestamp for comparison.

1.  Modify the `LogWritable` class to implement the `org.apache.hadoop.io.WritableComparable` interface:

```
public class LogWritable implements
   WritableComparable<LogWritable> {

   private Text userIP, timestamp, request;
   private IntWritable responseSize, status;

   public LogWritable() {
     this.userIP = new Text();
     this.timestamp=  new Text();
     this.request = new Text();
     this.responseSize = new IntWritable();
     this.status = new IntWritable();
   }

   public void readFields(DataInput in) throws IOException {
     userIP.readFields(in);
     timestamp.readFields(in);
     request.readFields(in);
     responseSize.readFields(in);
     status.readFields(in);
   }
```

```
public void write(DataOutput out) throws IOException {
  userIP.write(out);
  timestamp.write(out);
  request.write(out);
  responseSize.write(out);
  status.write(out);
}

public int compareTo(LogWritable o) {
  if (userIP.compareTo(o.userIP)==0){
      return (timestamp.compareTo(o.timestamp));
  }else return (userIP.compareTo(o.userIP);
}

public boolean equals(Object o) {
  if (o instanceof LogWritable) {
      LogWritable other = (LogWritable) o;
      return userIP.equals(other.userIP)
        && timestamp.equals(other.timestamp);
  }
  return false;
}

public int hashCode()
{
  Return userIP.hashCode();
}
  ......... // getters and setters for the fields
}
```

2. You can use the `LogWritable` type as either a `key` type or a `value` type in your MapReduce computation. In the following example, we use the `LogWritable` type as the Map output `key` type:

```
public class LogProcessorMap extends Mapper<LongWritable,
Text, LogWritable, IntWritable> {
...
}

public class LogProcessorReduce extends
  Reducer<LogWritable,
IntWritable, Text, IntWritable> {
```

```
public void reduce(LogWritablekey,
Iterable<IntWritable> values, Context context) {
      ……  }
}
```

3. Configure the output types of the job accordingly.

```
Job job = ……
...
job.setOutputKeyClass(Text.class);
job.setOutputValueClass(IntWritable.class);
job.setMapOutputKeyClass(LogWritable.class);
job.setMapOutputValueClass(IntWritable.class);
```

## How it works...

The `WritableComparable` interface introduces the `compareTo()` method in addition to the `readFields()` and `write()` methods of the `Writable` interface. The `compareTo()` method should return a negative integer, zero, or a positive integer, if this object is less than, equal to, or greater than the object being compared to it respectively. In the `LogWritable` implementation, we consider the objects equal if both user's IP addresses and the timestamps are the same. If the objects are not equal, we decide the sort order, first based on the user IP address and then based on the timestamp.

```
public int compareTo(LogWritable o) {
   if (userIP.compareTo(o.userIP)==0){
       return (timestamp.compareTo(o.timestamp));
   }else return (userIP.compareTo(o.userIP);
}
```

Hadoop uses `HashPartitioner` as the default partitioner implementation to calculate the distribution of the intermediate data to the Reducers. `HashPartitioner` requires the `hashCode()` method of the key objects to satisfy the following two properties:

▸ Provide the same hash value across different JVM instances

▸ Provide a uniform distribution of hash values

Hence, you must implement a stable `hashCode()` method for your custom Hadoop `key` types satisfying both the earlier-mentioned requirements. In the `LogWritable` implementation, we use the hash code of the request hostname/IP address as the hash code of the `LogWritable` instance. This ensures that the intermediate `LogWritable` data will be partitioned based on the request hostname/IP address.

```
public int hashCode()
{
   return userIP.hashCode();
}
```

## See also

The *Implementing a custom Hadoop Writable data type* recipe.

# Emitting data of different value types from a Mapper

Emitting data products belonging to multiple value types from a Mapper is useful when performing Reducer-side joins as well as when we need to avoid the complexity of having multiple MapReduce computations to summarize different types of properties in a dataset. However, Hadoop Reducers do not allow multiple input value types. In these scenarios, we can use the `GenericWritable` class to wrap multiple `value` instances belonging to different data types.

In this recipe, we reuse the HTTP server log entry analyzing the sample of the *Implementing a custom Hadoop Writable data type* recipe. However, instead of using a custom data type, in the current recipe, we output multiple value types from the Mapper. This sample aggregates the total number of bytes served from the web server to a particular host and also outputs a tab-separated list of URLs requested by the particular host. We use `IntWritable` to output the number of bytes from the Mapper and `Text` to output the request URL.

## How to do it...

The following steps show how to implement a Hadoop `GenericWritable` data type that can wrap instances of either `IntWritable` or `Text` data types:

1. Write a class extending the `org.apache.hadoop.io.GenericWritable` class. Implement the `getTypes()` method to return an array of the `Writable` classes that you will be using. If you are adding a custom constructor, make sure to add a parameter-less default constructor as well.

```
public class MultiValueWritable extends GenericWritable {

  private static Class[] CLASSES =  new Class[]{
    IntWritable.class,
    Text.class
  };

  public MultiValueWritable(){
  }

  public MultiValueWritable(Writable value){
```

```
        set(value);
    }

    protected Class[] getTypes() {
      return CLASSES;
    }
}
```

2. Set `MultiValueWritable` as the output value type of the Mapper. Wrap the output `Writable` values of the Mapper with instances of the `MultiValueWritable` class.

```
public class LogProcessorMap extends
    Mapper<Object, Text, Text, MultiValueWritable> {
  private Text userHostText = new Text();
  private Text requestText = new Text();
  private IntWritable responseSize = new IntWritable();

  public void map(Object key, Text value,
                              Context context)…{
    ……// parse the value (log entry) using a regex.
    userHostText.set(userHost);
    requestText.set(request);
    bytesWritable.set(responseSize);

    context.write(userHostText,
    new MultiValueWritable(requestText));
    context.write(userHostText,
    new MultiValueWritable(responseSize));
  }
}
```

3. Set the Reducer input value type as `MultiValueWritable`. Implement the `reduce()` method to handle multiple value types.

```
public class LogProcessorReduce extends
  Reducer<Text,MultiValueWritable,Text,Text> {
  private Text result = new Text();

  public void reduce(Text key,
  Iterable<MultiValueWritable>values, Context context)…{
  int sum = 0;
  StringBuilder requests = new StringBuilder();
  for (MultiValueWritable multiValueWritable : values) {
    Writable writable = multiValueWritable.get();
    if (writable instanceof IntWritable){
```

```
      sum += ((IntWritable)writable).get();
    }else{
      requests.append(((Text)writable).toString());
      requests.append("\t");
    }
  }
}
  result.set(sum + "\t"+requests);
  context.write(key, result);
  }
}
```

4. Set `MultiValueWritable` as the Map output value class of this computation:

```
    Job job = …
    job.setMapOutputValueClass(MultiValueWritable.class);
```

## How it works...

The `GenericWritable` implementations should extend `org.apache.hadoop.`
`io.GenericWritable` and should specify a set of the `Writable` value types to wrap, by
returning an array of `CLASSES` from the `getTypes()` method. The `GenericWritable`
implementations serialize and de-serialize the data using the index to this array of classes.

```
    private static Class[] CLASSES =  new Class[]{
      IntWritable.class,
      Text.class
    };

    protected Class[] getTypes() {
      return CLASSES;
    }
```

In the Mapper, you wrap each of your values with instances of the `GenericWritable`
implementation:

```
    private Text requestText = new Text();
    context.write(userHostText,
    new MultiValueWritable(requestText));
```

The Reducer implementation has to take care of the different value types manually.

```
    if (writable instanceof IntWritable){
      sum += ((IntWritable)writable).get();
    }else{
      requests.append(((Text)writable).toString());
      requests.append("\t");
    }
```

## There's more...

`org.apache.hadoop.io.ObjectWritable` is another class that can be used to achieve the same objective as `GenericWritable`. The `ObjectWritable` class can handle Java primitive types, strings, and arrays without the need of a `Writable` wrapper. However, Hadoop serializes the `ObjectWritable` instances by writing the class name of the instance with each serialized entry, making it inefficient compared to a `GenericWritable` class-based implementation.

## See also

The *Implementing a custom Hadoop Writable data type* recipe.

# Choosing a suitable Hadoop InputFormat for your input data format

Hadoop supports processing of many different formats and types of data through InputFormat. The InputFormat of a Hadoop MapReduce computation generates the key-value pair inputs for the Mappers by parsing the input data. InputFormat also performs the splitting of the input data into logical partitions, essentially determining the number of Map tasks of a MapReduce computation and indirectly deciding the execution location of the Map tasks. Hadoop generates a Map task for each logical data partition and invokes the respective Mappers with the key-value pairs of the logical splits as the input.

## How to do it...

The following steps show you how to use `FileInputFormat` based `KeyValueTextInputFormat` as InputFormat for a Hadoop MapReduce computation:

1. In this example, we are going to specify the `KeyValueTextInputFormat` as InputFormat for a Hadoop MapReduce computation using the `Job` object as follows:

```
Configuration conf = new Configuration();
Job job = new Job(conf, "log-analysis");
……
job.SetInputFormatClass(KeyValueTextInputFormat.class)
```

2. Set the input paths to the job:

```
FileInputFormat.setInputPaths(job, new Path(inputPath));
```

## How it works...

`KeyValueTextInputFormat` is an input format for plain text files, which generates a key-value record for each line of the input text files. Each line of the input data is broken into a key (text) and value (text) pair using a delimiter character. The default delimiter is the tab character. If a line does not contain the delimiter, the whole line will be treated as the key and the value will be empty. We can specify a custom delimiter by setting a property in the job's configuration object as follows, where we use the comma character as the delimiter between the key and value.

```
conf.set("key.value.separator.in.input.line", ",");
```

`KeyValueTextInputFormat` is based on `FileInputFormat`, which is the base class for the file-based InputFormats. Hence, we specify the input path to the MapReduce computation using the `setInputPaths()` method of the `FileInputFormat` class. We have to perform this step when using any InputFormat that is based on the `FileInputFormat` class.

```
FileInputFormat.setInputPaths(job, new Path(inputPath));
```

We can provide multiple HDFS input paths to a MapReduce computation by providing a comma-separated list of paths. You can also use the `addInputPath()` static method of the `FileInputFormat` class to add additional input paths to a computation.

```
public static void setInputPaths(JobConf conf,Path... inputPaths)
public static void addInputPath(JobConf conf, Path path)
```

## There's more...

Make sure that your Mapper input data types match the data types generated by InputFormat used by the MapReduce computation.

The following are some of the InputFormat implementations that Hadoop provides to support several common data formats:

► `TextInputFormat`: This is used for plain text files. `TextInputFormat` generates a key-value record for each line of the input text files. For each line, the key (`LongWritable`) is the byte offset of the line in the file and the value (`Text`) is the line of text. `TextInputFormat` is the default InputFormat of Hadoop.

▶ NLineInputFormat: This is used for plain text files. NLineInputFormat splits the input files into logical splits of fixed numbers of lines. We can use NLineInputFormat when we want our Map tasks to receive a fixed number of lines as the input. The key (LongWritable) and value (Text) records are generated for each line in the split similar to the TextInputFormat class. By default, NLineInputFormat creates a logical split (and a Map task) per line. The number of lines per split (or key-value records per Map task) can be specified as follows. NLineInputFormat generates a key-value record for each line of the input text files.

```
NLineInputFormat.setNumLinesPerSplit(job,50);
```

▶ SequenceFileInputFormat: This is used for Hadoop SequenceFile input data. Hadoop SequenceFiles store the data as binary key-value pairs and support data compression. SequenceFileInputFormat is useful when using the result of a previous MapReduce computation in SequenceFile format as the input of a MapReduce computation. The following are its subclasses:

  ❏ SequenceFileAsBinaryInputFormat: This is a subclass of the SequenceInputFormat class that presents the key (BytesWritable) and the value (BytesWritable) pairs in raw binary format.

  ❏ SequenceFileAsTextInputFormat: This is a subclass of the SequenceInputFormat class that presents the key (Text) and the value (Text) pairs as strings.

▶ DBInputFormat: This supports reading the input data for MapReduce computation from a SQL table. DBInputFormat uses the record number as the key (LongWritable) and the query result record as the value (DBWritable).

## See also

The *Adding support for new input data formats – implementing a custom InputFormat* recipe

# Adding support for new input data formats – implementing a custom InputFormat

Hadoop enables us to implement and specify custom InputFormat implementations for our MapReduce computations. We can implement custom InputFormat implementations to gain more control over the input data as well as to support proprietary or application-specific input data file formats as inputs to Hadoop MapReduce computations. An InputFormat implementation should extend the org.apache.hadoop.mapreduce. InputFormat<K,V> abstract class overriding the createRecordReader() and getSplits() methods.

In this recipe, we implement an InputFormat and a RecordReader for the HTTP log files. This InputFormat will generate `LongWritable` instances as keys and `LogWritable` instances as the values.

## How to do it...

The following are the steps to implement a custom InputFormat for the HTTP server log files based on the `FileInputFormat` class:

1. `LogFileInputFormat` operates on the data in HDFS files. Hence, we implement the `LogFileInputFormat` subclass extending the `FileInputFormat` class:

```
public class LogFileInputFormat extends
  FileInputFormat<LongWritable, LogWritable>{

  public RecordReader<LongWritable, LogWritable>
  createRecordReader(InputSplit arg0,
  TaskAttemptContext arg1) throws …… {
    return new LogFileRecordReader();
  }

}
```

2. Implement the `LogFileRecordReader` class:

```
public class LogFileRecordReader extends
RecordReader<LongWritable, LogWritable>{

  LineRecordReader lineReader;
  LogWritable value;

  public void initialize(InputSplit inputSplit,
  TaskAttemptContext attempt)…{
    lineReader = new LineRecordReader();
    lineReader.initialize(inputSplit, attempt);
  }

  public boolean nextKeyValue() throws IOException, ..{
    if (!lineReader.nextKeyValue()){
      return false;
  }

    String line  =lineReader.getCurrentValue().toString();
    …………//Extract the fields from 'line' using a regex
```

```
        value = new LogWritable(userIP, timestamp, request,
            status, bytes);
        return true;
    }

    public LongWritable getCurrentKey() throws..{
        return lineReader.getCurrentKey();
    }

    public LogWritable getCurrentValue() throws ..{
        return value;
    }

    public float getProgress() throws IOException ..{
        return lineReader.getProgress();
    }

    public void close() throws IOException {
        lineReader.close();
    }
}
```

3. Specify `LogFileInputFormat` as InputFormat for the MapReduce computation using the `Job` object as follows. Specify the input paths for the computations using the underlying `FileInputFormat`.

```
Job job = ……
……
job.setInputFormatClass(LogFileInputFormat.class);
FileInputFormat.setInputPaths(job, new Path(inputPath));
```

4. Make sure the Mappers of the computation use `LongWritable` as the input key type and `LogWritable` as the input `value` type:

```
public class LogProcessorMap extends
Mapper<LongWritable, LogWritable, Text, IntWritable>{
    public void map(LongWritable key,
        LogWritable value, Context context) throws ……{
        ………}
}
```

## How it works...

`LogFileInputFormat` extends `FileInputFormat`, which provides a generic splitting mechanism for HDFS-file based InputFormat. We override the `createRecordReader()` method in `LogFileInputFormat` to provide an instance of our custom `RecordReader` implementation, `LogFileRecordReader`. Optionally, we can also override the `isSplitable()` method of the `FileInputFormat` class to control whether the input files are split-up to logical partitions or used as whole files.

```
Public RecordReader<LongWritable, LogWritable>
  createRecordReader(InputSplit arg0,
  TaskAttemptContext arg1) throws …… {
    return new LogFileRecordReader();
}
```

The `LogFileRecordReader` class extends the `org.apache.hadoop.mapreduce.RecordReader<K,V>` abstract class and uses `LineRecordReader` internally to perform the basic parsing of the input data. `LineRecordReader` reads lines of text from the input data:

```
lineReader = new LineRecordReader();
lineReader.initialize(inputSplit, attempt);
```

We perform custom parsing of the log entries of the input data in the `nextKeyValue()` method. We use a regular expression to extract the fields out of the HTTP service log entry and populate an instance of the `LogWritable` class using those fields.

```
public boolean nextKeyValue() throws IOException, ..{
  if (!lineReader.nextKeyValue())
    return false;

  String line = lineReader.getCurrentValue().toString();
  …………//Extract the fields from 'line' using a regex

  value = new LogWritable(userIP, timestamp, request,
    status, bytes);
  return true;
}
```

## There's more...

We can perform custom splitting of input data by overriding the getSplits() method of the InputFormat class. The getSplits() method should return a list of InputSplit objects. An InputSplit object represents a logical partition of the input data and will be assigned to a single Map task. InputSplit classes extend the InputSplit abstract class and should override the getLocations() and getLength() methods. The getLength() method should provide the length of the split and the getLocations() method should provide a list of nodes where the data represented by this split is physically stored. Hadoop uses a list of data local nodes for Map task scheduling. The FileInputFormat class we use in the preceding example uses the org.apache.hadoop.mapreduce.lib.input.FileSplit as the InputSplit implementations.

You can write InputFormat implementations for non-HDFS data as well. The org.apache.hadoop.mapreduce.lib.db.DBInputFormat is one example of InputFormat. DBInputFormat supports reading the input data from a SQL table.

## See also

The *Choosing a suitable Hadoop InputFormat for your input data format* recipe.

# Formatting the results of MapReduce computations – using Hadoop OutputFormats

Often the output of your MapReduce computation will be consumed by other applications. Hence, it is important to store the result of a MapReduce computation in a format that can be consumed efficiently by the target application. It is also important to store and organize the data in a location that is efficiently accessible by your target application. We can use Hadoop OutputFormat interface to define the data storage format, data storage location, and the organization of the output data of a MapReduce computation. An OutputFormat prepares the output location and provides a RecordWriter implementation to perform the actual serialization and storage of data.

Hadoop uses the org.apache.hadoop.mapreduce.lib.output. TextOutputFormat<K,V> abstract class as the default OutputFormat for the MapReduce computations. TextOutputFormat writes the records of the output data to plain text files in HDFS using a separate line for each record. TextOutputFormat uses the tab character to delimit between the key and the value of a record. TextOutputFormat extends FileOutputFormat, which is the base class for all file-based output formats.

## How to do it...

The following steps show you how to use the `FileOutputFormat` based `SequenceFileOutputFormat` as the OutputFormat for a Hadoop MapReduce computation.

1.  In this example, we are going to specify the `org.apache.hadoop.mapreduce.lib.output.SequenceFileOutputFormat<K,V>` as the OutputFormat for a Hadoop MapReduce computation using the `Job` object as follows:

    ```
    Job job = ……
    ……
    job.setOutputFormatClass(SequenceFileOutputFormat.class)
    ```

2.  Set the output paths to the job:

    ```
    FileOutputFormat.setOutputPath(job, new Path(outputPath));
    ```

## How it works...

`SequenceFileOutputFormat` serializes the data to Hadoop SequenceFiles. Hadoop SequenceFiles store the data as binary key-value pairs and support data compression. SequenceFiles are efficient specially for storing non-text data. We can use the SequenceFiles to store the result of a MapReduce computation, if the output of the MapReduce computation is going to be the input of another Hadoop MapReduce computation.

`SequenceFileOutputFormat` is based on the `FileOutputFormat`, which is the base class for the file-based `OutputFormat`. Hence, we specify the output path to the MapReduce computation using the `setOutputPath()` method of the `FileOutputFormat`. We have to perform this step when using any OutputFormat that is based on the `FileOutputFormat`.

```
FileOutputFormat.setOutputPath(job, new Path(outputPath));
```

## There's more...

You can implement custom OutputFormat classes to write the output of your MapReduce computations in a proprietary or custom data format and/or to store the result in a storage other than HDFS by extending the `org.apache.hadoop.mapreduce.OutputFormat<K,V>` abstract class. In case your OutputFormat implementation stores the data in a filesystem, you can extend from the `FileOutputFormat` class to make your life easier.

# Writing multiple outputs from a MapReduce computation

We can use the `MultipleOutputs` feature of Hadoop to emit multiple outputs from a MapReduce computation. This feature is useful when we want to write different outputs to different files and also when we need to output an additional output in addition to the main output of a job. The `MultipleOutputs` feature allows us to specify a different OutputFormat for each output as well.

## How to do it...

The following steps show you how to use the `MultipleOutputs` feature to output two different datasets from a Hadoop MapReduce computation:

1. Configure and name the multiple outputs using the Hadoop driver program:

```
Job job = Job.getInstance(getConf(), "log-analysis");
...
FileOutputFormat.setOutputPath(job, new Path(outputPath));
MultipleOutputs.addNamedOutput(job, "responsesizes",
TextOutputFormat.class,Text.class, IntWritable.class);
MultipleOutputs.addNamedOutput(job, "timestamps",
TextOutputFormat.class,Text.class, Text.class);
```

2. Write data to the different named outputs from the `reduce` function:

```
public class LogProcessorReduce ...{
  private MultipleOutputs mos;

  protected void setup(Context context) .. {
    mos = new MultipleOutputs(context);
  }

  public void reduce(Text key, … {
    ...
    mos.write("timestamps", key, val.getTimestamp());
    ...
    mos.write("responsesizes", key, result);
  }
}
```

3. Close all the opened outputs by adding the following to the `cleanup` function of the Reduce task:

```
@Override
  public void cleanup(Context context) throws IOException,
    InterruptedException {
      mos.close();
  }
```

4. Output filenames will be in the format `namedoutput-r-xxxxx` for each output type written. For the current sample, example output filenames would be `responsesizes-r-00000` and `timestamps-r-00000`.

## How it works...

We first add the named outputs to the job in the driver program using the following static method of the `MultipleOutputs` class:

```
public static addNamedOutput(Job job, String namedOutput, Class<?
extends OutputFormat> outputFormatClass, Class<?> keyClass,
Class<?> valueClass)
```

Then we initialize the `MultipleOutputs` feature in the `setup` method of the Reduce task as follows:

```
protected void setup(Context context) .. {
  mos = new MultipleOutputs(context);
  }
```

We can write the different outputs using the names we defined in the driver program using the following method of the `MultipleOutputs` class:

```
public <K,V> void write (String namedOutput, K key, V value)
```

You can directly write to an output path without defining the named outputs using the following method of the `MultipleOutputs`. This output will use the OutputFormat defined for the job to format the output.

```
public void write(KEYOUT key, VALUEOUT value,
  String baseOutputPath)
```

Finally, we make sure to close all the outputs from the `cleanup` method of the Reduce task using the close method of the `MultipleOutputs` class. This should be done to avoid loss of any data written to the different outputs.

```
public void close()
```

## Using multiple input data types and multiple Mapper implementations in a single MapReduce application

We can use the `MultipleInputs` feature of Hadoop to run a MapReduce job with multiple input paths, while specifying a different InputFormat and (optionally) a Mapper for each path. Hadoop will route the outputs of the different Mappers to the instances of the single Reducer implementation of the MapReduce computation. Multiple inputs with different InputFormats are useful when we want to process multiple datasets with the same meaning but different InputFormats (comma-delimited dataset and tab-delimited dataset).

We can use the following `addInputPath` static method of the `MutlipleInputs` class to add the input paths and the respective InputFormats to the MapReduce computation:

```
Public static void addInputPath(Job job, Path path,
   Class<?extends InputFormat>inputFormatClass)
```

The following is an example usage of the preceding method:

```
MultipleInputs.addInputPath(job, path1, CSVInputFormat.class);
MultipleInputs.addInputPath(job, path1, TabInputFormat.class);
```

Multiple inputs feature with the ability to specify different Mapper implementations and `InputFormats` is useful when performing a Reduce-side join of two or more datasets:

```
public static void addInputPath(JobConfconf, Path path,
      Class<?extends InputFormat>inputFormatClass,
      Class<?extends Mapper>mapperClass)
```

The following is an example of using multiple inputs with different InputFormats and different Mapper implementations:

```
MultipleInputs.addInputPath(job, accessLogPath,
   TextInputFormat.class, AccessLogMapper.class);
MultipleInputs.addInputPath(job, userDataPath,
   TextInputFormat.class, UserDataMapper.class);
```

### See also

The *Adding support for new input data formats – implementing a custom InputFormat* recipe.

# Hadoop intermediate data partitioning

Hadoop MapReduce partitions the intermediate data generated by the *Map tasks across the Reduce tasks* of the computations. A proper partitioning function ensuring balanced load for each Reduce task is crucial to the performance of MapReduce computations. Partitioning can also be used to group together related sets of records to specific reduce tasks, where you want certain outputs to be processed or grouped together. The figure in the *Introduction* section of this chapter depicts where the partitioning fits into the overall MapReduce computation flow.

Hadoop partitions the intermediate data based on the key space of the intermediate data and decides which Reduce task will receive which intermediate record. The sorted set of keys and their values of a partition would be the input for a Reduce task. In Hadoop, the total number of partitions should be equal to the number of Reduce tasks for the MapReduce computation. Hadoop partitioners should extend the `org.apache.hadoop.mapreduce.Partitioner<KEY,VALUE>` abstract class. Hadoop uses `org.apache.hadoop.mapreduce.lib.partition.HashPartitioner` as the default partitioner for the MapReduce computations. **HashPartitioner** partitions the keys based on their `hashcode()`, using the formula *key.hashcode() mod r*, where *r* is the number of Reduce tasks. The following diagram illustrates `HashPartitioner` for a computation with two Reduce tasks:

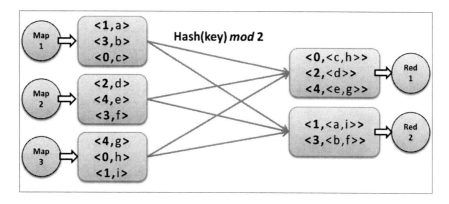

There can be scenarios where our computations logic would require or can be better implemented using an application's specific data-partitioning schema. In this recipe, we implement a custom partitioner for our HTTP log processing application, which partitions the keys (IP addresses) based on their geographic regions.

## How to do it...

The following steps show you how to implement a custom partitioner that partitions the intermediate data based on the location of the request IP address or the hostname:

1.  Implement the `IPBasedPartitioner` class extending the `Partitioner` abstract class:

    ```
    public class IPBasedPartitioner extends Partitioner
      <Text, IntWritable>{

      public int getPartition(Text ipAddress,
        IntWritable value, int numPartitions) {
        String region = getGeoLocation(ipAddress);

        if (region!=null){
          return ((region.hashCode() &
            Integer.MAX_VALUE) % numPartitions);
        }
      return 0;
      }
    }
    ```

2.  Set the `Partitioner` class parameter in the `Job` object:

    ```
    Job job = ......
    ......
    job.setPartitionerClass(IPBasedPartitioner.class);
    ```

## How it works...

In the preceding example, we perform the partitioning of the intermediate data, such that the requests from the same geographic region will be sent to the same Reducer instance. The `getGeoLocation()` method returns the geographic location of the given IP address. We omit the implementation details of the `getGeoLocation()` method as it's not essential for the understanding of this example. We then obtain the `hashCode()` method of the geographic location and perform a modulo operation to choose the Reducer bucket for the request.

## There's more...

`TotalOrderPartitioner` and `KeyFieldPartitioner` are two of the several built-in partitioner implementations provided by Hadoop.

## TotalOrderPartitioner

`TotalOrderPartitioner` extends `org.apache.hadoop.mapreduce.lib.partition.TotalOrderPartitioner<K,V>`. The set of input records to a Reducer are in a sorted order ensuring proper ordering within an input partition. However, the Hadoop default partitioning strategy (`HashPartitioner`) does not enforce an ordering when partitioning the intermediate data and scatters the keys among the partitions. In use cases where we want to ensure a global order, we can use the `TotalOrderPartitioner` to enforce a total order to reduce the input records across the Reducer task. `TotalOrderPartitioner` requires a partition file as the input defining the ranges of the partitions. The `org.apache.hadoop.mapreduce.lib.partition.InputSampler` utility allows us to generate a partition file for the `TotalOrderPartitioner` by sampling the input data. `TotalOrderPartitioner` is used in the Hadoop TeraSort benchmark.

```
Job job = ……
……
job.setPartitionerClass(TotalOrderPartitioner.class);
TotalOrderPartitioner.setPartitionFile(jobConf,partitionFile);
```

## KeyFieldBasedPartitioner

The `org.apache.hadoop.mapreduce.lib.partition.KeyFieldBasedPartitioner<K,V>` abstract class can be used to partition the intermediate data based on parts of the key. A key can be split into a set of fields by using a separator string. We can specify the indexes of the set of fields to be considered when partitioning. We can also specify the index of the characters within fields.

# Secondary sorting – sorting Reduce input values

MapReduce frameworks sort the Reduce input data based on the key of the key-value pairs and also group the data based on the key. Hadoop invokes the `reduce` function for each unique key in the sorted order of keys with the list of values belonging to that key as the second parameter. However, the list of values for each key is not sorted in any particular order. There are many scenarios where it would be useful to have the list of Reduce input values for each key sorted based on some criteria as well. The examples include finding the maximum or minimum value for a given key without iterating the whole list, to optimize Reduce-side joins, to identify duplicate data products, and so on.

We can use the Hadoop framework to sort the Reduce input values using a mechanism called secondary sorting. We achieve this by forcing Hadoop framework to sort the reduce input key-value pairs using the key as well as using several designated fields from the value. However, the partitioning of Map output data and the grouping of the Reduce input data is still performed only using the key. This assures that the Reduce input data is grouped and sorted by the key, while the list of values belonging to a key would be in a sorted order as well.

## How to do it...

The following steps show you how to perform secondary sorting of the Reduce input values in a Hadoop MapReduce computation:

1. First, implement a `WritableComparable` data type that can contain the actual key (`visitorAddress`) and the fields (`responseSize`) from the value that needs to be included in the sorted order. The comparator for this new compound type should enforce the sorting order, where the actual key comes first, followed by the sorting criteria derived from the value fields contained in this new type. We use this compound type as the Map output key type. Alternatively, you can also use an existing `WritableComparable` type as the Map output key type, which contains the actual key and the other fields from the value, by providing a comparable implementation for that data type to enforce the sorting order.

```
public class SecondarySortWritable … {
  private String visitorAddress;
  private int responseSize;
  .........
  @Override
  public boolean equals(Object right) {
    if (right instanceof SecondarySortWritable) {
      SecondarySortWritable r = (SecondarySortWritable)
      right;
      return (r.visitorAddress.equals(visitorAddress) &&
      (r.responseSize == responseSize));
    } else {
      return false;
    }
  }

  @Override
  public int compareTo(SecondarySortWritable o) {
    if (visitorAddress.equals(o.visitorAddress)) {
      return responseSize < o.responseSize ? -1 : 1;
    } else {
      return visitorAddress.compareTo(o.visitorAddress);
    }
  }
}
```

2. Modify the `map` and `reduce` functions to use the compound key that we created:

```
public class LogProcessorMap extends Mapper<Object,
  LogWritable, SecondarySortWritable, LogWritable > {
  private SecondarySortWritable outKey ..

  public void map(Object key, …..{
    outKey.set(value.getUserIP().toString(),
    value.getResponseSize().get());
    context.write(outKey,value);
  }
}

public class LogProcessorReduce extends
  Reducer<SecondarySortWritable, LogWritable..{
  ……
  public void reduce(SecondarySortWritable key,
  Iterable<LogWritable> values {
    ……
  }
}
```

3. Implement a custom partitioner to partition the Map output data based only on the actual key (`visitorAddress`) contained in the compound key:

```
public class SingleFieldPartitioner extends… {
  public int getPartition(SecondarySortWritable key,
  Writable value, int numPartitions) {
    return (int)(key.getVisitorAddress().hashCode() %
    numPartitions);
  }
}
```

4. Implement a custom grouping comparator to group the Reduce inputs only based on the actual key (`visitorAddress`):

```
public class GroupingComparator extends WritableComparator
{
  public GroupingComparator() {
    super(SecondarySortWritable.class, true);
  }

  @Override
  public int compare(WritableComparable o1,
  WritableComparable o2) {
```

```
        SecondarySortWritable firstKey =
        (SecondarySortWritable) o1;
        SecondarySortWritable secondKey =
        (SecondarySortWritable) o2;
        return (firstKey.getVisitorAddress())
        .compareTo(secondKey.getVisitorAddress());
    }
```

5. Configure the partitioner, `GroupingComparator`, and the Map output key type in the driver program:

```
Job job = Job.getInstance(getConf(), "log-analysis");
......
job.setMapOutputKeyClass(SecondarySortWritable.class);
......

// group and partition by the visitor address
job.setPartitionerClass(SingleFieldPartitioner.class);
job.setGroupingComparatorClass(GroupingComparator.class);
```

## How it works...

We first implemented a custom `WritableComparable` key type that would hold the actual key and the sort fields of the value. We ensure the sorting order of this new compound key type to be the actual key followed by the sort fields from the value. This will ensure that the Reduce input data would be first sorted based on the actual key followed by the given fields of the value.

Then we implemented a custom partitioner that would partition the Map output data only based on the actual key field from the new compound key. This step ensures that each key-value pair with the same actual key would be processed by the same Reducer. Finally, we implemented a grouping comparator that would consider only the actual key field of the new key when grouping the reduced input key-value pairs. This ensures that each `reduce` function input will be the new compound key together with the list of values belonging to the actual key. The list of values would be in sorted order as that is defined in the comparator of the compound key.

## See also

The *Adding support for new input data formats – implementing a custom InputFormat* recipe.

# Broadcasting and distributing shared resources to tasks in a MapReduce job – Hadoop DistributedCache

We can use the Hadoop **DistributedCache** to distribute read-only file-based resources to the Map and Reduce tasks. These resources can be simple data files, archives, or JAR files that are needed for the computations performed by the Mappers or the Reducers.

## How to do it...

The following steps show you how to add a file to the Hadoop DistributedCache and how to retrieve it from the Map and Reduce tasks:

1. Copy the resource to the HDFS. You can also use files that are already there in the HDFS.

   ```
   $ hadoop fs -copyFromLocal ip2loc.dat ip2loc.dat
   ```

2. Add the resource to the DistributedCache from your driver program:

   ```
   Job job = Job.getInstance……
   ……
   job.addCacheFile(new URI("ip2loc.dat#ip2location"));
   ```

3. Retrieve the resource in the setup() method of your Mapper or Reducer and use the data in the map() or reduce() function:

   ```
   public class LogProcessorMap extends
     Mapper<Object, LogWritable, Text, IntWritable> {
     private IPLookup lookupTable;

     public void setup(Context context) throws IOException{

       File lookupDb = new File("ip2location");
       // Load the IP lookup table (a simple hashmap of ip
       // prefixes as keys and country names as values) to
       // memory
       lookupTable = IPLookup.LoadData(lookupDb);
     }

     public void map(…) {
       String country =
         lookupTable.getCountry(value.ipAddress);
         ……
     }
   }
   ```

## How it works...

Hadoop copies the files added to the DistributedCache to all the worker nodes before the execution of any task of the job. DistributedCache copies these files only once per job. Hadoop also supports creating symlinks to the DistributedCache files in the working directory of the computation by adding a fragment with the desired symlink name to the URI. In the following example, we are using `ip2location` as the symlink to the `ip2loc.dat` file in the DistributedCache:

```
job.addCacheArchive(new URI("/data/ip2loc.dat#ip2location"));
```

We parse and load the data from the DistributedCache in the `setup()` method of the Mapper or the Reducer. Files with symlinks are accessible from the working directory using the provided symlink's name.

```
private IPLookup lookup;
public void setup(Context context) throws IOException{

  File lookupDb = new File("ip2location");
  // Load the IP lookup table to memory
  lookup = IPLookup.LoadData(lookupDb);
}

public void map(…) {
  String location =lookup.getGeoLocation(value.ipAddress);
  ……
}
```

We can also access the data in the DistributedCache directly using the `getLocalCacheFiles()` method, without using the symlink:

```
URI[] cacheFiles = context.getCacheArchives();
```

 DistributedCache does not work in Hadoop local mode.

## There's more...

The following sections show you how to distribute the compressed archives using DistributedCache, how to add resources to the DistributedCache using the command line, and how to use the DistributedCache to add resources to the classpath of the Mapper and the Reducer.

## Distributing archives using the DistributedCache

We can use the DistributedCache to distribute archives as well. Hadoop extracts the archives in the worker nodes. You can also provide symlinks to the archives using the URI fragments. In the next example, we use the `ip2locationdb` symlink for the `ip2locationdb.tar.gz` archive.

Consider the following MapReduce driver program:

```
Job job = ......
job.addCacheArchive(
  new URI("/data/ip2locationdb.tar.gz#ip2locationdb"));
```

The extracted directory of the archive can be accessible from the working directory of the Mapper or the Reducer using the symlink provided earlier:

Consider the following Mapper program:

```
public void setup(Context context) throws IOException{
  Configuration conf = context.getConfiguration();

  File lookupDbDir = new File("ip2locationdb");
  String[] children = lookupDbDir.list();

  ...
}
```

You can also access the non-extracted DistributedCache archived files directly using the following method in the Mapper or Reducer implementation:

```
URI[] cachePath;

public void setup(Context context) throws IOException{
  Configuration conf = context.getConfiguration();
  cachePath = context.getCacheArchives();
  ...
}
```

## Adding resources to the DistributedCache from the command line

Hadoop supports adding files or archives to the DistributedCache using the command line, provided that your MapReduce driver programs implement the `org.apache.hadoop.util.Tool` interface or utilize `org.apache.hadoop.util.GenericOptionsParser`. Files can be added to the DistributedCache using the `-files` command-line option, while archives can be added using the `-archives` command-line option. Files or archives can be in any filesystem accessible for Hadoop, including your local filesystem.

These options support a comma-separated list of paths and the creation of symlinks using the URI fragments.

```
$ hadoop jar C4LogProcessor.jar LogProcessor
  -files ip2location.dat#ip2location  indir outdir
$ hadoop jar C4LogProcessor.jar LogProcessor
  -archives ip2locationdb.tar.gz#ip2locationdb indir outdir
```

### Adding resources to the classpath using the DistributedCache

You can use DistributedCache to distribute JAR files and other dependent libraries to the Mapper or Reducer. You can use the following methods in your driver program to add the JAR files to the classpath of the JVM running the Mapper or the Reducer:

```
public static void addFileToClassPath(
   Path file,Configuration conf,FileSystem fs)

public static void addArchiveToClassPath(
   Path archive,Configuration conf, FileSystem fs)
```

Similar to the -files and -archives command-line options we described in the *Adding resources to the DistributedCache from the command line* subsection, we can also add the JAR files to the classpath of our MapReduce computations by using the -libjars command-line option. In order for the -libjars command-line option to work, MapReduce driver programs should implement the Tool interface or should utilize GenericOptionsParser.

```
$ hadoop jar C4LogProcessor.jar LogProcessor
  -libjars ip2LocationResolver.jar  indir outdir
```

# Using Hadoop with legacy applications – Hadoop streaming

Hadoop streaming allows us to use any executable or a script as the Mapper or the Reducer of a Hadoop MapReduce job. Hadoop streaming enables us to perform rapid prototyping of the MapReduce computations using Linux shell utility programs or using scripting languages. Hadoop streaming also allows the users with some or no Java knowledge to utilize Hadoop to process data stored in HDFS.

In this recipe, we implement a Mapper for our HTTP log processing application using Python and use a Hadoop aggregate-package-based Reducer.

## How to do it...

The following are the steps to use a Python program as the Mapper to process the HTTP
server log files:

1. Write the `logProcessor.py` python script:

```python
#!/usr/bin/python
import sys
import re
def main(argv):
    regex =re.compile('......')
    line = sys.stdin.readline()
    try:
      while line:
        fields = regex.match(line)
        if(fields!=None):
          print"LongValueSum:"+fields.group(1)+
            "\t"+fields.group(7)
        line =sys.stdin.readline()
    except"end of file":
      return None
if __name__ == "__main__":
    main(sys.argv)
```

2. Use the following command from the Hadoop installation directory to execute the
   Streaming MapReduce computation:

```
$ hadoop jar \
    $HADOOP_MAPREDUCE_HOME/hadoop-streaming-*.jar \
    -input indir \
    -output outdir \
    -mapper logProcessor.py \
    -reducer aggregate \
    -file logProcessor.py
```

## How it works...

Each Map task launches the Hadoop streaming executable as a separate process in
the worker nodes. The input records (the entries or lines of the log file, not broken into
key-value pairs) to the Mapper are provided as lines to the standard input of that process.
The executable should read and process the records from the standard input until the end
of the file is reached.

```python
line = sys.stdin.readline()
try:
```

```
      while line:
        .........
        line =sys.stdin.readline()
  except "end of file":
      return None
```

Hadoop streaming collects the outputs of the executable from the standard output of the process. Hadoop streaming converts each line of the standard output to a key-value pair, where the text up to the first tab character is considered the key and the rest of the line as the value. The `logProcessor.py` python script outputs the key-value pairs, according to this convention, as follows:

```
If (fields!=None):
      print "LongValueSum:"+fields.group(1)+ "\t"+fields.group(7);
```

In our example, we use the Hadoop `aggregate` package for the reduction part of our computation. The Hadoop `aggregate` package provides Reducer and combiner implementations for simple aggregate operations such as sum, max, unique value count, and histogram. When used with Hadoop streaming, the Mapper outputs must specify the type of aggregation operation of the current computation as a prefix to the output key, which is the `LongValueSum` in our example.

Hadoop streaming also supports the distribution of files to the worker nodes using the `-file` option. We can use this option to distribute executable files, scripts, or any other auxiliary file needed for the streaming computation. We can specify multiple `-file` options for a computation.

```
$ hadoop jar …… \
   -mapper logProcessor.py \
   -reducer aggregate \
   -file logProcessor.py
```

## There's more...

We can specify Java classes as the Mapper and/or Reducer and/or combiner programs of Hadoop streaming computations. We can also specify InputFormat and other options to a Hadoop streaming computation.

Hadoop streaming also allows us to use Linux shell utility programs as Mapper and Reducer. The following example shows the usage of `grep` as the Mapper of a Hadoop streaming computation.

```
$ hadoop jar
   $HADOOP_MAPREDUCE_HOME/hadoop-streaming-*.jar \
   -input indir \
   -output outdir \
   -mapper 'grep "wiki"'
```

Hadoop streaming provides the Reducer input records of each key group line by line to the standard input of the process that is executing the executable. However, Hadoop streaming does not have a mechanism to distinguish when it starts to feed records of a new key to the process. Hence, the scripts or the executables for Reducer programs should keep track of the last seen key of the input records to demarcate between key groups.

Extensive documentation on Hadoop streaming is available at `http://hadoop.apache.org/mapreduce/docs/stable1/streaming.html`.

## See also

The *Data preprocessing using Hadoop streaming and Python* and *De-duplicating data using Hadoop streaming* recipes in *Chapter 10, Mass Text Data Processing*.

# Adding dependencies between MapReduce jobs

Often we require multiple MapReduce applications to be executed in a workflow-like manner to achieve our objective. Hadoop `ControlledJob` and `JobControl` classes provide a mechanism to execute a simple workflow graph of MapReduce jobs by specifying the dependencies between them.

In this recipe, we execute the `log-grep` MapReduce computation followed by the `log-analysis` MapReduce computation on an HTTP server log dataset. The `log-grep` computation filters the input data based on a regular expression. The `log-analysis` computation analyses the filtered data. Hence, the `log-analysis` computation is dependent on the `log-grep` computation. We use the `ControlledJob` class to express this dependency and use the `JobControl` class to execute both the related MapReduce computations.

## How to do it...

The following steps show you how to add a MapReduce computation as a dependency of another MapReduce computation:

1.  Create the `Configuration` and the `Job` objects for the first MapReduce job and populate them with the other needed configurations:

    ```
    Job job1 = ……
    job1.setJarByClass(RegexMapper.class);
    job1.setMapperClass(RegexMapper.class);
    FileInputFormat.setInputPaths(job1, new Path(inputPath));
    FileOutputFormat.setOutputPath(job1, new
      Path(intermedPath));
    ……
    ```

2. Create the `Configuration` and `Job` objects for the second MapReduce job and populate them with the necessary configurations:

```
Job job2 = ......
job2.setJarByClass(LogProcessorMap.class);
job2.setMapperClass(LogProcessorMap.class);
job2.setReducerClass(LogProcessorReduce.class);
FileOutputFormat.setOutputPath(job2, new Path(outputPath));
.........
```

3. Set the output directory of the first job as the input directory of the second job:

```
FileInputFormat.setInputPaths
    (job2, new Path(intermedPath +"/part*"));
```

4. Create `ControlledJob` objects using the `Job` objects created earlier:

```
ControlledJob controlledJob1 =
    new ControlledJob(job1.getConfiguration());
ControlledJob controlledJob2 =
    new ControlledJob(job2.getConfiguration());
```

5. Add the first job as a dependency to the second job:

```
controlledJob2.addDependingJob(controlledJob1);
```

6. Create the `JobControl` object for this group of jobs and add the `ControlledJob` objects created in step 4 to the newly created `JobControl` object:

```
JobControl jobControl = new
    JobControl("JobControlDemoGroup");
jobControl.addJob(controlledJob1);
jobControl.addJob(controlledJob2);
```

7. Create a new thread to run the group of jobs added to the `JobControl` object. Start the thread and wait for its completion:

```
Thread jobControlThread = new Thread(jobControl);
jobControlThread.start();
while (!jobControl.allFinished()){
  Thread.sleep(500);
}
jobControl.stop();
```

## How it works...

The `ControlledJob` class encapsulates the MapReduce job and keeps track of the job's dependencies. A `ControlledJob` class with depending jobs becomes ready for submission only when all of its depending jobs are completed successfully. A `ControlledJob` class fails if any of the depending jobs fail.

The `JobControl` class encapsulates a set of `ControlledJobs` and their dependencies. `JobControl` tracks the status of the encapsulated `ControlledJobs` and contains a thread that submits the jobs that are in the *READY* state.

If you want to use the output of a MapReduce job as the input of a dependent job, the input paths to the dependent job have to be set manually. By default, Hadoop generates an output folder per Reduce task name with the `part` prefix. We can specify all the `part` prefixed subdirectories as input to the dependent job using wildcards.

```
FileInputFormat.setInputPaths
    (job2, new Path(job1OutPath +"/part*"));
```

## There's more...

We can use the `JobControl` class to execute and track a group of non-dependent tasks as well.

Apache **Oozie** is a workflow system for Hadoop MapReduce computations. You can use Oozie to execute **Directed Acyclic Graphs** (**DAG**) of MapReduce computations. You can find more information on Oozie from the project's home page at `http://oozie.apache.org/`.

The `ChainMapper` class, available in the older version of Hadoop MapReduce API, allowed us to execute a pipeline of Mapper classes inside a single Map task computation in a pipeline. `ChainReducer` provided similar support for Reduce tasks. This API still exists in Hadoop 2 for backward compatibility reasons.

# Hadoop counters to report custom metrics

Hadoop uses a set of counters to aggregate the metrics for MapReduce computations. Hadoop counters are helpful to understand the behavior of our MapReduce programs and to track the progress of the MapReduce computations. We can define custom counters to track the application-specific metrics in MapReduce computations.

## How to do it...

The following steps show you how to define a custom counter to count the number of bad or corrupted records in our log processing application:

1. Define the list of custom counters using `enum`:

```
public static enum LOG_PROCESSOR_COUNTER {
   BAD_RECORDS
   };
```

2. Increment the counter in your Mapper or Reducer:

```
context.getCounter(LOG_PROCESSOR_COUNTER.BAD_RECORDS)
.increment(1);
```

3. Add the following to your driver program to access the counters:

```
Job job = new Job(getConf(), "log-analysis");
......
Counters counters = job.getCounters();
Counter badRecordsCounter =
counters.findCounter(LOG_PROCESSOR_COUNTER.BAD_RECORDS);
System.out.println("# of Bad Records:"+
   badRecordsCounter.getValue());
```

4. Execute your Hadoop MapReduce computation. You can also view the counter values in the admin console or in the command line.

```
$ hadoop jar C4LogProcessor.jar \
   demo.LogProcessor in out 1

.........

12/07/29 23:59:01 INFO mapred.JobClient: Job complete:
job_201207271742_0020

12/07/29 23:59:01 INFO mapred.JobClient: Counters: 30

12/07/29 23:59:01 INFO mapred.JobClient:
demo.LogProcessorMap$LOG_PROCESSOR_COUNTER

12/07/29 23:59:01 INFO mapred.JobClient:    BAD_RECORDS=1406

12/07/29 23:59:01 INFO mapred.JobClient:    Job Counters

.........

12/07/29 23:59:01 INFO mapred.JobClient:    Map output
records=112349

# of Bad Records :1406
```

## How it works...

You have to define your custom counters using `enum`. The set of counters in an `enum` will form a group of counters. The ApplicationMaster aggregates the counter values reported by the Mappers and the Reducers.

# 5
# Analytics

In this chapter, we will cover the following recipes:

- ▸ Simple analytics using MapReduce
- ▸ Performing GROUP BY using MapReduce
- ▸ Calculating frequency distributions and sorting using MapReduce
- ▸ Plotting the Hadoop MapReduce results using gnuplot
- ▸ Calculating histograms using MapReduce
- ▸ Calculating Scatter plots using MapReduce
- ▸ Parsing a complex dataset with Hadoop
- ▸ Joining two datasets using MapReduce

## Introduction

In this chapter, we will discuss how we can use Hadoop to process a dataset and to understand its basic characteristics. We will cover more complex methods like data mining, classification, clustering, and so on, in later chapters.

This chapter will show how you can calculate basic analytics using a given dataset. For the recipes in this chapter, we will use two datasets:

- ▸ The NASA weblog dataset available at `http://ita.ee.lbl.gov/html/contrib/NASA-HTTP.html` is a real-life dataset collected using the requests received by NASA web servers. You can find a description of the structure of this data at this link. A small extract of this dataset that can be used for testing is available inside the `chapter5/resources` folder of the code repository.

- ▸ List of e-mail archives of Apache Tomcat developers available from `http://tomcat.apache.org/mail/dev/`. These archives are in the MBOX format.

The contents of this chapter are based on the *Chapter 6, Analytics,* of the previous edition of this book, Hadoop MapReduce Cookbook. That chapter was contributed by the coauthor Srinath Perera.

**Sample code**

The example code files for this book are available in GitHub at `https://github.com/thilg/hcb-v2`. The `chapter5` folder of the code repository contains the sample source code files for this chapter.

Sample codes can be compiled by issuing the `gradle build` command in the `chapter5` folder of the code repository. Project files for Eclipse IDE can be generated by running the `gradle eclipse` command in the main folder of the code repository. Project files for IntelliJ IDEA IDE can be generated by running the `gradle idea` command in the main folder of the code repository.

# Simple analytics using MapReduce

Aggregate metrics such as mean, max, min, standard deviation, and so on, provide the basic overview of a dataset. You may perform these calculations, either for the whole dataset or to a subset or a sample of the dataset.

In this recipe, we will use Hadoop MapReduce to calculate the minimum, maximum, and average size of files served from a web server, by processing logs of the web server. The following figure shows the execution flow of this computation:

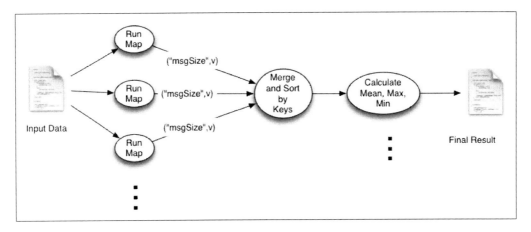

As shown in the figure, the **Map** function emits the size of the file as the value and the string `msgSize` as the key. We use a single Reduce task, and all the intermediate key-value pairs will be sent to that Reduce task. Then, the **Reduce** function calculates the aggregate values using the information emitted by the Map tasks.

## Getting ready

This recipe assumes that you have a basic understanding of how the processing of Hadoop MapReduce works. If not, please follow the *Writing a WordCount MapReduce application, bundling it, and running it using the Hadoop local mode* and *Setting up Hadoop YARN in a distributed cluster environment using Hadoop v2* recipes from *Chapter 1, Getting Started with Hadoop v2*. You need to have a working Hadoop installation as well.

## How to do it...

The following steps describe how to use MapReduce to calculate simple aggregate metrics about the weblog dataset:

1.  Download the weblog dataset from `ftp://ita.ee.lbl.gov/traces/NASA_access_log_Jul95.gz` and extract it. Let's call the extracted location as `<DATA_DIR>`.

2.  Upload the extracted data to HDFS by running the following commands:
    ```
    $ hdfs dfs -mkdir data
    $ hdfs dfs -mkdir data/weblogs
    $ hdfs dfs -copyFromLocal \
    <DATA_DIR>/NASA_access_log_Jul95 \
    data/weblogs
    ```

3.  Compile the sample source code for this chapter by running the `gradle build` command from the `chapter5` folder of the source repository.

4.  Run the MapReduce job by using the following command:
    ```
    $ hadoop jar hcb-c5-samples.jar \
    chapter5.weblog.MsgSizeAggregateMapReduce \
    data/weblogs data/msgsize-out
    ```

5.  Read the results by running the following command:
    ```
    $ hdfs dfs -cat data/msgsize-out/part*
    ….
    Mean      1150
    Max       6823936
    Min       0
    ```

## How it works...

You can find the source file for this recipe from `chapter5/src/chapter5/weblog/MsgSizeAggregateMapReduce.java`.

HTTP logs follow a standard pattern as follows. The last token is the size of the web page served:

```
205.212.115.106 - - [01/Jul/1995:00:00:12 -0400] "GET
/shuttle/countdown/countdown.html HTTP/1.0" 200 3985
```

We will use the Java regular expressions to parse the log lines, and the `Pattern.compile()` method at the top of the class defines the regular expression. Regular expressions are a very useful tool while writing text-processing Hadoop computations:

```
public void map(Object key, Text value, Context context) … {
  Matcher matcher = httplogPattern.matcher(value.toString());
  if (matcher.matches()) {
    int size = Integer.parseInt(matcher.group(5));
    context.write(new Text("msgSize"), new IntWritable(size));
  }
}
```

Map tasks receive each line in the log file as a different key-value pair. It parses the lines using regular expressions and emits the file size as the value with `msgSize` as the key.

Then, Hadoop collects all the output key-value pairs from the Map tasks and invokes the Reduce task. Reducer iterates all the values and calculates the minimum, maximum, and mean size of the files served from the web server. It is worth noting that by making the values available as an iterator, Hadoop allows us to process the data without storing them in memory, allowing the Reducers to scale to large datasets. Whenever possible, you should process the `reduce` function input values without storing them in memory:

```
public void reduce(Text key, Iterable<IntWritable> values, …{
  double total = 0;
  int count = 0;
  int min = Integer.MAX_VALUE;
  int max = 0;

  Iterator<IntWritable> iterator = values.iterator();
  while (iterator.hasNext()) {
    int value = iterator.next().get();
    total = total + value;
    count++;
```

```
    if (value < min)
      min = value;

    if (value > max)
      max = value;
  }
  context.write(new Text("Mean"),
    new IntWritable((int) total / count));
  context.write(new Text("Max"), new IntWritable(max));
  context.write(new Text("Min"), new IntWritable(min));
}
```

The `main()` method of the job looks similar to the WordCount example, except for the highlighted lines that have been changed to accommodate the output data types:

```
job.setOutputKeyClass(Text.class);
job.setOutputValueClass(IntWritable.class);
```

## There's more...

You can learn more about Java regular expressions from the Java tutorial, `http://docs.oracle.com/javase/tutorial/essential/regex/`.

# Performing GROUP BY using MapReduce

This recipe shows how we can use MapReduce to group data into simple groups and calculate metrics for each group. We will use the web server's log dataset for this recipe as well. This computation is similar to the `select page, count(*) from weblog_table group by page` SQL statement. The following figure shows the execution flow of this computation:

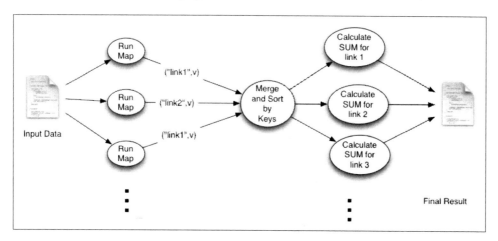

As shown in the figure, the **Map** tasks emit the requested URL path as the key. Then, Hadoop sorts and groups the intermediate data by the key. All values for a given key will be provided into a single Reduce function invocation, which will count the number of occurrences of that URL path.

## Getting ready

This recipe assumes that you have a basic understanding of how Hadoop MapReduce processing works.

## How to do it...

The following steps show how we can group web server log data and calculate analytics:

1. Download the weblog dataset from `ftp://ita.ee.lbl.gov/traces/NASA_` `access_log_Jul95.gz` and extract it. Let's call the extracted location as `<DATA_DIR>`

2. Upload the extracted data to HDFS by running the following commands:

   ```
   $ hdfs dfs -mkdir data
   $ hdfs dfs -mkdir data/weblogs
   $ hdfs dfs -copyFromLocal \
   <DATA_DIR>/NASA_access_log_Jul95 \
   data/weblogs
   ```

3. Compile the sample source code for this chapter by running the `gradle build` command from the `chapter5` folder of the source repository.

4. Run the MapReduce job using the following command:

   ```
   $ hadoop jar hcb-c5-samples.jar \
   chapter5.weblog.HitCountMapReduce \
   data/weblogs data/hit-count-out
   ```

5. Read the results by running the following command:

   ```
   $ hdfs dfs -cat data/hit-count-out/part*
   ```

You will see that it will print the results as follows:

```
/base-ops/procurement/procurement.html   28
/biomed/                                  1
/biomed/bibliography/biblio.html          7
/biomed/climate/airqual.html              4
/biomed/climate/climate.html              5
/biomed/climate/gif/f16pcfinmed.gif       4
/biomed/climate/gif/f22pcfinmed.gif       3
/biomed/climate/gif/f23pcfinmed.gif       3
/biomed/climate/gif/ozonehrlyfin.gif      3
```

## How it works...

You can find the source for this recipe from `chapter5/src/chapter5/HitCountMapReduce.java`.

As described in the earlier recipe, we will use a regular expression to parse the web server logs and to extract the requested URL paths. For example, `/shuttle/countdown/countdown.html` will get extracted from the following sample log entry:

```
205.212.115.106 - - [01/Jul/1995:00:00:12 -0400] "GET
/shuttle/countdown/countdown.html HTTP/1.0" 200 3985
```

The following code segment shows the Mapper:

```java
private final static IntWritable one = new IntWritable(1);
private Text word = new Text();
public void map(Object key, Text value, Context context) …… {
  Matcher matcher = httplogPattern.matcher(value.toString());
  if (matcher.matches()) {
    String linkUrl = matcher.group(4);
    word.set(linkUrl);
    context.write(word, one);
  }
}
```

Map tasks receive each line in the log file as a different key-value pair. Map tasks parse the lines using regular expressions and emit the link as the key and number `one` as the value.

Then, Hadoop collects all values for different keys (links) and invokes the Reducer once for each link. Then, each Reducer counts the number of hits for each link:

```
private IntWritable result = new IntWritable();
public void reduce(Text key, Iterable<IntWritable> values,… {
  int sum = 0;
  for (IntWritable val : values) {
    sum += val.get();
  }
  result.set(sum);
  context.write(key, result);
}
```

# Calculating frequency distributions and sorting using MapReduce

**Frequency distribution** is the number of hits received by each URL sorted in ascending order. We already calculated the number of hits for each URL in the earlier recipe. This recipe will sort that list based on the number of hits.

## Getting ready

This recipe assumes that you have a working Hadoop installation. This recipe will use the results from the *Performing GROUP BY using MapReduce* recipe of this chapter. Follow this recipe if you have not done so already.

## How to do it...

The following steps show how to calculate frequency distribution using MapReduce:

1. Run the MapReduce job using the following command. We assume that the `data/hit-count-out` path contains the output of the `HitCountMapReduce` computation of the previous recipe:

   ```
   $ bin/hadoop jar hcb-c5-samples.jar \
   chapter5.weblog.FrequencyDistributionMapReduce \
   data/hit-count-out data/freq-dist-out
   ```

2. Read the results by running the following command:

   ```
   $ hdfs dfs -cat data/freq-dist-out/part*
   ```

You will see that it will print the results similar to the following:

```
/cgi-bin/imagemap/countdown?91,175        12
/cgi-bin/imagemap/countdown?105,143       13
/cgi-bin/imagemap/countdown70?177,284     14
```

## How it works...

The *Performing GROUP BY using MapReduce* recipe of this chapter calculates the number of hits received by each URL path. MapReduce sorts the Map output's intermediate key-value pairs by their keys before invoking the `reduce` function. In this recipe, we use this sorting feature to sort the data based on the number of hits.

You can find the source for this recipe from `chapter5/src/chapter5/FrequencyDistributionMapReduce.java`.

The Map task outputs the number of hits as the key and the URL path as the value:

```
public void map(Object key, Text value, Context context) ...... {
   String[] tokens = value.toString().split("\\s");
   context.write(new IntWritable(Integer.parseInt(tokens[1])),
   new Text(tokens[0]));
}
```

The Reduce task receives the key-value pairs sorted by the key (number of hits):

```
public void reduce(IntWritable key, Iterable<Text> values, ...... {
   Iterator<Text> iterator = values.iterator();
   while (iterator.hasNext()) {
   context.write(iterator.next(), key);
   }
}
```

We use only a single Reduce task in this computation in order to ensure a global ordering of the results.

## There's more...

It's possible to achieve a global ordering even with multiple Reduce tasks, by utilizing the Hadoop `TotalOrderPartitioner`. Refer to the *Hadoop intermediate data partitioning* recipe of *Chapter 4*, *Developing Complex Hadoop MapReduce Applications*, for more information on the `TotalOrderPartitioner`.

# Plotting the Hadoop MapReduce results using gnuplot

Although Hadoop MapReduce jobs can generate interesting analytics, making sense of those results and getting a detailed understanding about the data often requires us to see the overall trends in the data. The human eye is remarkably good at detecting patterns, and plotting the data often yields a deeper understanding of the data. Therefore, we often plot the results of Hadoop jobs using a plotting program.

This recipe explains how to use gnuplot, which is a free and powerful plotting program used to plot Hadoop results.

## Getting ready

This recipe assumes that you have followed the previous recipe, *Calculating frequency distributions and sorting using MapReduce*. If you have not done this, follow this recipe. Install the gnuplot plotting program by following the instructions in `http://www.gnuplot.info/`.

## How to do it...

The following steps show how to plot Hadoop job results using gnuplot:

1. Download the results of the previous recipe to a local computer by running the following command:

   ```
   $ hdfs dfs -copyToLocal data/freq-dist-out/part-r-00000 2.data
   ```

2. Copy all the `*.plot` files from the `chapter5/plots` folder to the location of the downloaded data.

3. Generate the plot by running the following command:

   ```
   $ gnuplot httpfreqdist.plot
   ```

4. It will generate a file called `freqdist.png`, which will look like the following:

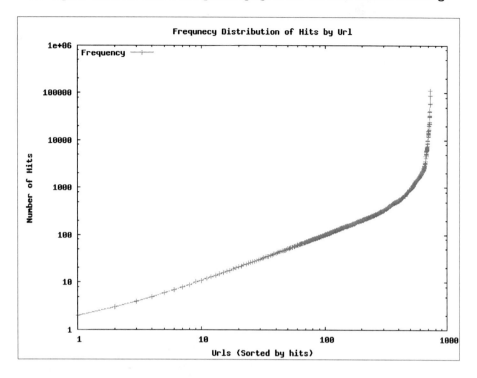

The preceding plot is plotted in log-log scale, and the first part of the distribution follows the **zipf** (power law) distribution, which is a common distribution seen in the Web. The last few most popular links have much higher rates than expected from a zipf distribution.

Discussion about more details on this distribution is out of the scope of this book. However, this plot demonstrates the kind of insights we can get by plotting the analytical results. In most of the future recipes, we will use gnuplot to plot and analyze the results.

## How it works...

The following steps describe how plotting with gnuplot works:

> ▶ You can find the source for the gnuplot file from `chapter5/plots/`
> `httpfreqdist.plot`. The source for the plot will look like the following:
>
> ```
> set terminal png
> set output "freqdist.png"
>
> set title "Frequnecy Distribution of Hits by Url";
> set ylabel "Number of Hits";
> set xlabel "Urls (Sorted by hits)";
> set key left top
> set log y
> set log x
>
> plot"2.data" using 2 title "Frequency" with linespoints
> ```

> ▶ Here, the first two lines define the output format. This example uses PNG, but the gnuplot supports many other terminals such as SCREEN, PDF, EPS, and so on.

> ▶ The next four lines define the axis labels and the title.

> ▶ The next two lines define the scale of each axis, and this plot uses log scale for both.

> ▶ The last line defines the plot. Here, it is asking gnuplot to read the data from the `2.data` file, and use the data in the second column of the file via `using 2` and to plot it using lines. Columns must be separated by whitespaces.

> ▶ Here, if you want to plot one column against another, for example, data from column 1 against column 2, you should write `using 1:2` instead of `using 2`.

## There's more...

You can learn more about gnuplot from `http://www.gnuplot.info/`.

# Calculating histograms using MapReduce

Another interesting view of a dataset is a **histogram**. A histogram makes sense only under a continuous dimension (for example, accessed time and file size). It groups the number of occurrences of an event into several groups in the dimension. For example, in this recipe, if we take the accessed time as the dimension, then we will group the accessed time by the hour.

The following figure shows the execution summary of this computation. The Mapper emits the hour of the access as the key and **1** as the value. Then, each `reduce` function invocation receives all the occurrences of a certain hour of the day, and it calculates the total number of occurrences for that hour of the day.

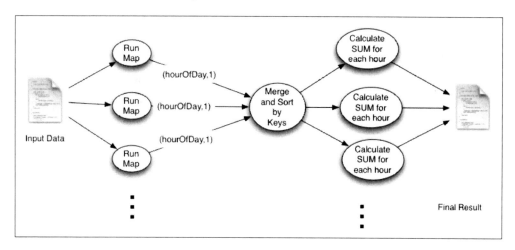

## Getting ready

This recipe assumes that you have a working Hadoop installation. Install gnuplot.

## How to do it...

The following steps show how to calculate and plot a histogram:

1. Download the weblog dataset from `ftp://ita.ee.lbl.gov/traces/NASA_access_log_Jul95.gz` and extract it.

2. Upload the extracted data to HDFS by running the following commands:

   ```
   $ hdfs dfs -mkdir data
   $ hdfs dfs -mkdir data/weblogs
   $ hdfs dfs -copyFromLocal \
   <DATA_DIR>/NASA_access_log_Jul95 \
   data/weblogs
   ```

3. Compile the sample source code for this chapter by running the `gradle build` command from the `chapter5` folder of the source repository.

4. Run the MapReduce job using the following command:

```
$ hadoop jar hcb-c5-samples.jar \
chapter5.weblog.HistogramGenerationMapReduce \
data/weblogs data/histogram-out
```

5. Inspect the results by running the following command:

```
$ hdfs dfs -cat data/histogram-out/part*
```

6. Download the results to a local computer by running the following command:

```
$ hdfs dfs -copyToLocal data/histogram-out/part-r-00000 3.data
```

7. Copy all the `*.plot` files from the `chapter5/plots` folder to the location of the downloaded data.

8. Generate the plot by running the following command:

```
$gnuplot httphistbyhour.plot
```

9. It will generate a file called `hitsbyHour.png`, which will look like the following:

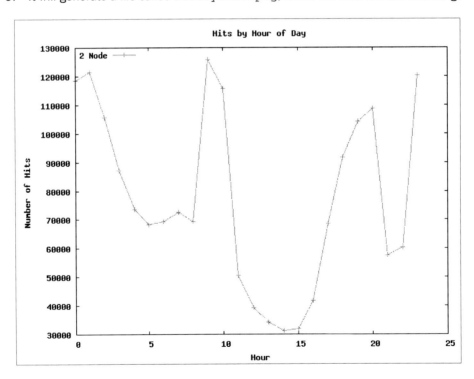

## How it works...

You can find the source code for this recipe from `chapter5/src/chapter5/weblog/HistogramGenerationMapReduce.java`. Similar to the earlier recipes of this chapter, we use a regular expression to parse the log file and extract the access time from the log files.

The following code segment shows the `map` function:

```
public void map(Object key, Text value, Context context) ... {
  try {
    Matcher matcher = httplogPattern.matcher(value.toString());
    if (matcher.matches()) {
      String timeAsStr = matcher.group(2);
      Date time = dateFormatter.parse(timeAsStr);
      Calendar calendar = GregorianCalendar.getInstance();
      calendar.setTime(time);
      int hour = calendar.get(Calendar.HOUR_OF_DAY);
      context.write(new IntWritable(hour), one);
    }
  } ......
}
```

The `map` function extracts the access time for each web page access and extracts the hour of the day from the access time. It emits the hour of the day as the key and `one` as the value.

Then, Hadoop collects all key-value pairs, sorts them, and then invokes the Reduce function once for each key. Reduce tasks calculate the total page views for each hour:

```
public void reduce(IntWritable key, Iterable<IntWritable>
values,..{
  int sum = 0;
  for (IntWritable val : values) {
    sum += val.get();
  }
  context.write(key, new IntWritable(sum));
}
```

# Calculating Scatter plots using MapReduce

Another useful tool while analyzing data is a **Scatter plot**, which can be used to find the relationship between two measurements (dimensions). It plots the two dimensions against each other.

For example, this recipe analyzes the data to find the relationship between the size of the web pages and the number of hits received by the web page.

The following image shows the execution summary of this computation. Here, the `map` function calculates and emits the message size (rounded to 1024 bytes) as the key and `one` as the value. Then, the Reducer calculates the number of occurrences for each message size:

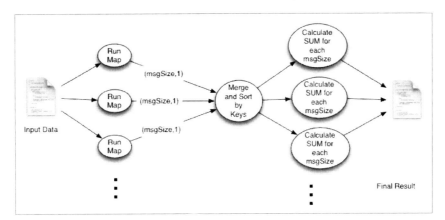

## Getting ready

This recipe assumes that you have a working Hadoop installation. Install gnuplot.

## How to do it...

The following steps show how to use MapReduce to calculate the correlation between two datasets:

1. Download the weblog dataset from `ftp://ita.ee.lbl.gov/traces/NASA_access_log_Jul95.gz` and extract it.

2. Upload the extracted data to HDFS by running the following commands:

```
$ hdfs dfs -mkdir data
$ hdfs dfs -mkdir data/weblogs
$ hdfs dfs -copyFromLocal \
<DATA_DIR>/NASA_access_log_Jul95 \
data/weblogs
```

3. Compile the sample source code for this chapter by running the `gradle build` command from the `chapter5` folder of the source repository.

4. Run the MapReduce job using the following command:

```
$ hadoop jar hcb-c5-samples.jar \
chapter5.weblog.MsgSizeScatterMapReduce \
data/weblogs data/scatter-out
```

5. Inspect the results by running the following command:

```
$ hdfs dfs -cat data/scatter-out/part*
```

6. Download the results of the previous recipe to the local computer by running the following command from HADOOP_HOME:

```
$ hdfs dfs -copyToLocal data/scatter-out/part-r-00000 5.data
```

7. Copy all the *.plot files from the chapter5/plots folder to the location of the downloaded data.

8. Generate the plot by running the following command:

```
$ gnuplot httphitsvsmsgsize.plot
```

9. It will generate a file called hitsbymsgSize.png, which will look like the following image:

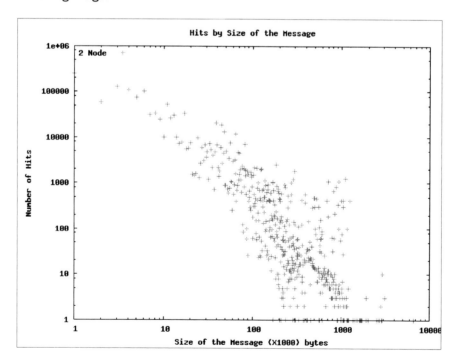

The plot shows a negative correlation between the number of hits and the size of the messages in the log scales.

## How it works...

You can find the source for the recipe from `chapter5/src/chapter5/MsgSizeScatterMapReduce.java`.

The following code segment shows the `map` function:

```
public void map(Object key, Text value, Context context) ......
{
  Matcher matcher =
  httplogPattern.matcher(value.toString());
  if (matcher.matches()) {
    int size = Integer.parseInt(matcher.group(5));
    context.write(new IntWritable(size / 1024), one);
  }
}
```

Map tasks parse the log entries and emit the file size in kilobytes as the key and `one` as the value.

Each Reducer walks through the values and calculates the count of page accesses for each file size:

```
public void reduce(IntWritable key, Iterable<IntWritable>
values,......{
  int sum = 0;
  for (IntWritable val : values) {
    sum += val.get();
  }
  context.write(key, new IntWritable(sum));
}
```

# Parsing a complex dataset with Hadoop

The datasets we used so far contained a data item in a single line, making it possible for us to use Hadoop default parsing support to parse those datasets. However, some datasets have more complex formats, where a single data item may span multiple lines. In this recipe, we will analyze mailing list archives of Tomcat developers. In the archive, each e-mail consists of multiple lines of the archive file. Therefore, we will write a custom Hadoop InputFormat to process the e-mail archive.

This recipe parses the complex e-mail list archives, and finds the owner (the person who started the thread) and the number of replies received by each e-mail thread.

The following figure shows the execution summary of this computation. The **Map** function emits the subject of the mail as the key, and the sender's e-mail address combined with the date as the value. Then, Hadoop groups the data by the e-mail subject and sends all the data related to that thread to the same Reducer.

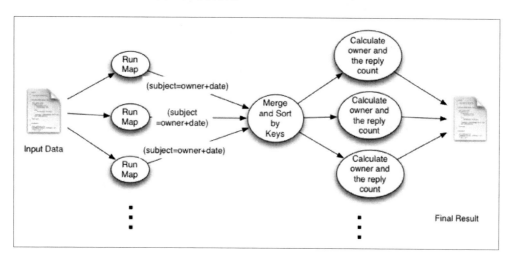

Then, the Reduce tasks identify the creator of each e-mail thread and the number of replies received by each thread.

## Getting ready

This recipe assumes that you have a working Hadoop installation.

## How to do it...

The following steps describe how to parse the Tomcat e-mail list dataset that has complex data format using Hadoop by writing an input formatter:

1. Download and extract the Apache Tomcat developer list e-mail archives for the year 2012 from `http://tomcat.apache.org/mail/dev/`. We call the destination folder as `DATA_DIR`.

2. Upload the extracted data to HDFS by running the following commands:

```
$ hdfs dfs -mkdir data
$ hdfs dfs -mkdir data/mbox
$ hdfs dfs -copyFromLocal \
<DATA_DIR>/* \
data/mbox
```

3. Compile the sample source code for this chapter by running the `gradle build` command from the `chapter5` folder of the source repository.

4. Run the MapReduce job using the following command:

```
$ hadoop jar hcb-c5-samples.jar \
chapter5.mbox.CountReceivedRepliesMapReduce \
data/mbox data/count-replies-out
```

5. Inspect the results by running the following command:

```
$ hdfs dfs -cat data/count-replies-out/part*
```

## How it works...

As explained before, this dataset has data items that span multiple lines. Therefore, we have to write a custom InputFormat and a custom RecordReader to parse the data. Source code files for this recipe are the `CountReceivedRepliesMapReduce.java`, `MBoxFileInputFormat.java`, and `MBoxFileReader.java` files in the `chapter5/src/chapter5/mbox` directory of the source code archive.

We add the new InputFormat to the Hadoop job via the Hadoop driver program as highlighted in the following code snippet:

```
Job job = Job.getInstance(getConf(), "MLReceiveReplyProcessor");
job.setJarByClass(CountReceivedRepliesMapReduce.class);
job.setMapperClass(AMapper.class);
job.setReducerClass(AReducer.class);
job.setNumReduceTasks(numReduce);

job.setOutputKeyClass(Text.class);
job.setOutputValueClass(Text.class);
job.setInputFormatClass(MBoxFileInputFormat.class);
FileInputFormat.setInputPaths(job, new Path(inputPath));
FileOutputFormat.setOutputPath(job, new Path(outputPath));

int exitStatus = job.waitForCompletion(true) ? 0 : 1;
```

As shown in the following code, the new formatter creates a RecordReader, which is used by Hadoop to read the key-value pair input to the Map tasks:

```
public class MboxFileFormat extends FileInputFormat<Text, Text>{
  private MBoxFileReaderboxFileReader = null;
  public RecordReader<Text, Text> createRecordReader(
  InputSplit inputSplit, TaskAttemptContext attempt) …{
    fileReader = new MBoxFileReader();
```

```
        fileReader.initialize(inputSplit, attempt);
        return fileReader;
    }
}
```

The following code snippets show the functionality of the RecordReader:

```
public class MBoxFileReader extends RecordReader<Text, Text> {

    public void initialize(InputSplitinputSplit, … {
      Path path = ((FileSplit) inputSplit).getPath();
      FileSystem fs = FileSystem.get(attempt.getConfiguration());
      FSDataInputStream fsStream = fs.open(path);
      reader = new BufferedReader(new InputStreamReader(fsStream));
    }
    public Boolean nextKeyValue() ……{
      if (email == null) {
      return false;
    }
    count++;
    while ((line = reader.readLine()) != null) {
      Matcher matcher = pattern1.matcher(line);
      if (!matcher.matches()) {
        email.append(line).append("\n");
      } else {
        parseEmail(email.toString());
        email = new StringBuffer();
        email.append(line).append("\n");
        return true;
      }
    }
    parseEmail(email.toString());
    email = null;
    return true;
  }
    ………
```

The nextKeyValue() method of the RecordReader parses the file, and generates key-value pairs for the consumption by the Map tasks. Each value has the *from*, *subject*, and *time* of each e-mail separated by a # character.

The following code snippet shows the Map task source code:

```
public void map(Object key, Text value, Context context) …… {
  String[] tokens = value.toString().split("#");
  String from = tokens[0];
  String subject = tokens[1];
  String date = tokens[2].replaceAll(",", "");
  subject = subject.replaceAll("Re:", "");
  context.write(new Text(subject), new Text(date + "#" + from));
}
```

The Map task receives each e-mail in the archive files as a separate key-value pair. It parses the lines by breaking it by the #, and emits the `subject` as the key, and `time` and `from` as the value.

Then, Hadoop collects all key-value pairs, sorts them, and then invokes the Reducer once for each key. Since we use the e-mail subject as the key, each Reduce function invocation will receive all the information about a single e-mail thread. Then, the Reduce function will analyze all the e-mails of a thread and find out who sent the first e-mail and how many replies have been received by each e-mail thread as follows:

```
public void reduce(Text key, Iterable<Text> values, …{
  TreeMap<Long, String>replyData = new TreeMap<Long, String>();

  for (Text val : values) {
    String[] tokens = val.toString().split("#");
    if(tokens.length != 2)
    throw new IOException("Unexpected token "+ val.toString());

    String from = tokens[1];
    Date date = dateFormatter.parse(tokens[0]);
    replyData.put(date.getTime(), from);
  }

  String owner = replyData.get(replyData.firstKey());
  Int replyCount = replyData.size();

  Int selfReplies = 0;
  for(String from: replyData.values()){
    if(owner.equals(from)){
      selfReplies++;
    }
  }
  replyCount = replyCount - selfReplies;
  context.write(new Text(owner),
  new Text(replyCount+"#" + selfReplies));
}
```

## There's more...

Refer to the *Adding support for new input data formats – implementing a custom InputFormat* recipe of *Chapter 4, Developing Complex Hadoop MapReduce Applications*, for more information on implementing custom InputFormats.

# Joining two datasets using MapReduce

As we have already observed, Hadoop is very good at reading through a dataset and calculating the analytics. However, we will often have to merge two datasets to analyze the data. This recipe will explain how to join two datasets using Hadoop.

As an example, this recipe will use the Tomcat developer archives dataset. A common belief among the open source community is that the more a developer is involved with the community (for example, by replying to e-mail threads in the project's mailing lists and helping others and so on), the more quickly they will receive responses to their queries. In this recipe, we will test this hypothesis using the Tomcat developer mailing list.

To test this hypothesis, we will run the MapReduce jobs as explained in the following figure:

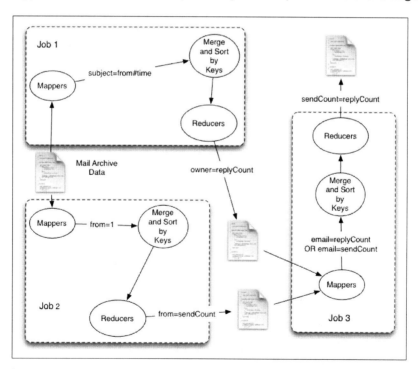

We will use the MBOX-formatted e-mail archives and use the custom InputFormat and RecordReader explained in the earlier recipe to parse them. Map tasks will receive the sender of the e-mail (from), the e-mail subject, and the time the e-mail was sent, as inputs.

1. In the first job, the `map` function will emit the subject as the key, and the sender's e-mail address and time as the value. Then, the Reducer step will receive all the values with the same subject and it will output the subject as the key, and the owner and reply count as the value. We executed this job in the previous recipe.

2. In the second job, the `map` function emits the sender's e-mail address as the key and `one` as the value. Then, the Reducer step will receive all the e-mails sent from the same address to the same Reducer. Using this data, each Reducer will emit the e-mail address as the key and the number of e-mails sent from that e-mail address as the value.

3. Finally, the third job reads both the outputs from the preceding two jobs, joins the results, and emits the number of e-mails sent by each e-mail address and the number of replies received by each e-mail address as the output.

## Getting ready

This recipe assumes that you have a working Hadoop installation. Follow the *Parsing a complex dataset with Hadoop* recipe. We will use the input data and the output data of that recipe in the following steps.

## How to do it...

The following steps show how to use MapReduce to join two datasets:

1. Run the `CountReceivedRepliesMapReduce` computation by following the *Parsing a complex dataset with Hadoop* recipe.

2. Run the second MapReduce job using the following command:

```
$ hadoop jar hcb-c5-samples.jar \
chapter5.mbox.CountSentRepliesMapReduce \
data/mbox data/count-emails-out
```

3. Inspect the results by using the following command:

```
$ hdfs dfs -cat data/count-emails-out/part*
```

4. Create a new folder `join-input` and copy both the results from the earlier jobs to that folder in HDFS:

```
$ hdfs dfs -mkdir data/join-input
$ hdfs dfs -cp \
```

```
data/count-replies-out/part-r-00000 \
data/join-input/1.data
$ hdfs dfs -cp \
data/count-emails-out/part-r-00000 \
data/join-input/2.data
```

5.  Run the third MapReduce job using the following command:

    ```
    $ hadoop jar hcb-c5-samples.jar \
    chapter5.mbox.JoinSentReceivedReplies \
    data/join-input data/join-out
    ```

6.  Download the results of step 5 to the local computer by running the following command:

    ```
    $ hdfs dfs -copyToLocal data/join-out/part-r-00000 8.data
    ```

7.  Copy all the `*.plot` files from the `chapter5/plots` folder to the location of the downloaded data.

8.  Generate the plot by running the following command:

    ```
    $ gnuplot sendvsreceive.plot
    ```

9.  It will generate a file called `sendreceive.png`, which will look like the following:

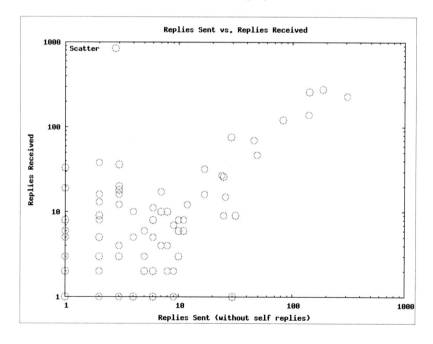

The graph confirms our hypothesis, and like before, the data approximately follows a power law distribution.

## How it works...

You can find the source code for this recipe from `chapter5/src/chapter5/mbox/CountSentRepliesMapReduce.java` and `chapter5/src/chapter5/mbox/JoinSentReceivedReplies.java`. We have already discussed the first job in the earlier recipe.

The following code snippet shows the `map` function for the second job. It receives the sender's e-mail, subject, and time separated by # as the input, which parses the input and outputs the sender's e-mail as the key, and the time the e-mail was sent as the value:

```
public void map(Object key, Text value, Context context) ......{
   String[] tokens = value.toString().split("#");
   String from = tokens[0];
   String date = tokens[2];
   context.write(new Text(from), new Text(date));
}
```

The following code snippet shows the `reduce` function for the second job. Each `reduce` function invocation receives the time of all the e-mails sent by one sender. The Reducer counts the number of replies sent by each sender, and outputs the sender's name as the key, and the number of replies sent as the value:

```
public void reduce(Text key, Iterable<Text> values, ......{
   int sum = 0;
   for (Text val : values) {
     sum = sum +1;
   }
   context.write(key, new IntWritable(sum));
}
```

The following code snippet shows the `map` function for the third job. It reads the outputs of the first and second jobs, and outputs them as the key-value pairs:

```
public void map(Object key, Text value, ...... {
   String[] tokens = value.toString().split("\\s");
   String from = tokens[0];
   String replyData = tokens[1];
   context.write(new Text(from), new Text(replyData));
}
```

The following code snippet shows the `reduce` function for the third job. Since both the outputs of the first and the second job have the same key, the number of replies sent and the number of replies received by a given user will be processed by the same Reducer. The `reduce` function removes self-replies and outputs the number of replies sent and the number of replies received as the key and value, thus joining the two datasets:

```
public void reduce(Text key, Iterable<Text> values, ...... {
    StringBuffer buf = new StringBuffer("[");
    try {
        int sendReplyCount = 0;
        int receiveReplyCount = 0;
        for (Text val : values) {
            String strVal = val.toString();
            if(strVal.contains("#")){
                String[] tokens = strVal.split("#");
                int repliesOnThisThread =Integer.parseInt(tokens[0]);
                int selfRepliesOnThisThread = Integer.parseInt(tokens[1]);
                receiveReplyCount = receiveReplyCount +
                repliesOnThisThread;
                sendReplyCount = sendReplyCount-selfRepliesOnThisThread;
            }else{
                sendReplyCount = sendReplyCount +
                Integer.parseInt(strVal);
            }
        }

        context.write(new IntWritable(sendReplyCount),
        new IntWritable(receiveReplyCount));
        buf.append("]");
    } ...
}
```

The final job is an example of using the MapReduce to join two datasets. The idea is to send all the values that need to be joined under the same key to the same Reducer, and join the data there.

# 6
# Hadoop Ecosystem – Apache Hive

In this chapter, we will cover the following recipes:

- ► Getting started with Apache Hive
- ► Creating databases and tables using Hive CLI
- ► Simple SQL-style data querying using Apache Hive
- ► Creating and populating Hive tables and views using Hive query results
- ► Utilizing different storage formats in Hive – storing table data using ORC files
- ► Using Hive built-in functions
- ► Hive batch mode – using a query file
- ► Performing a join with Hive
- ► Creating partitioned Hive tables
- ► Writing Hive User-defined Functions (UDF)
- ► HCatalog – performing Java MapReduce computations on data mapped to Hive tables
- ► HCatalog – Writing data to Hive tables from Java MapReduce computations

# Introduction

Hadoop has a family of projects that are either built on top of Hadoop or work very closely with Hadoop. These projects have given rise to an ecosystem that focuses on large-scale data processing, and often, users can use several of these projects in combination to solve their use cases. This chapter introduces Apache Hive, which provides data warehouse capabilities on top of the data stored in HDFS. *Chapter 7, Hadoop Ecosystem II – Pig, HBase, Mahout, and Sqoop* introduces a few other key projects in the Hadoop ecosystem.

Apache Hive provides an alternative high-level language layer to perform large-scale data analysis using Hadoop. Hive allows users to map the data stored in HDFS into tabular models and process them using HiveQL, the SQL-like language layer, to query very large datasets using Hadoop. HiveQL can be used to perform ad-hoc querying of datasets as well as for data summarizations and to perform data analytics. Due to its SQL-like language, Hive is a natural choice for users who are experienced with data warehousing using relational databases.

Hive translates the HiveQL queries to one or more MapReduce computations that perform the actual work. Hive allows us to define the structure on existing datasets using table schemas. However, Hive imposes this structure on the data only at the time of reading and processing the data (schema on read).

Hive is very good for analyzing very large datasets due to the large aggregate throughput it can achieve using the parallelism of the Hadoop cluster. However, Hive is not optimal for analyzing smaller datasets or for interactive queries due to the relatively high latencies of running MapReduce computations. Hive provides good scalability and fault tolerance capabilities through the use of MapReduce and HDFS underneath. Hive does not support transactions or row-level updates of data.

This chapter also introduces HCatalog, which is a component of Hive that provides a metadata abstraction layer for data stored in HDFS, making it easy for different components of the Hadoop ecosystem to process those data. HCatalog abstraction is based on a tabular model and augments structure, location, and other metadata information for the HDFS data. With HCatalog, we can use data processing tools such as Pig, Java MapReduce, and others without worrying about the structure, storage format, or the storage location of the data.

**Sample code**

The example code and data files for this book are available on GitHub at `https://github.com/thilg/hcb-v2`. The `chapter6` folder of the code repository contains the sample code for this chapter.

Sample codes can be compiled and built by issuing the `gradle build` command in the `chapter6` folder of the code repository. Project files for the Eclipse IDE can be generated by running the `gradle eclipse` command in the main folder of the code repository. Project files for IntelliJ IDEA IDE can be generated by running the `gradle idea` command in the main folder of the code repository.

In this chapter, we use the Book Crossing dataset as the sample data. This dataset is compiled by Cai-Nicolas Ziegler, and comprises a list of books, users, and ratings. The `Resources` folder of the source repository for this chapter contains a sample of the dataset. You can obtain the full dataset from `http://www2.informatik.uni-freiburg.de/~cziegler/BX/`.

# Getting started with Apache Hive

In order to install Hive, we recommend that you use a freely available commercial Hadoop distribution as described in *Chapter 1, Getting Started with Hadoop v2*. Another alternative is to use Apache Bigtop to install Hive. Refer to the Bigtop-related recipe in *Chapter 1, Getting Started with Hadoop v2* for steps on installation of Hive using the Apache Bigtop distribution.

## How to do it...

This section describes how to get started with Hive.

1. If you already have a working Hive installation, start the Hive **Command Line Interface** (**CLI**) by executing `hive` in the command prompt and skip to step 4:

   ```
   $ hive
   ```

2. In case you don't have a working Hive and Hadoop installation, the following couple of steps will guide you on how to install Hive with the MapReduce local mode. This is recommended only for learning and testing purposes. Download and extract the latest Hive version from `http://hive.apache.org/releases.html`:

   ```
   $ tar -zxvf hive-0.14.0.tar.gz
   ```

3. Start Hive by running the following commands from the extracted Hive folder:

   ```
   $ cd hive-0.14.0
   $ bin/hive
   ```

4. Optionally, you can set the following Hive property to enable Hive to print headers when displaying the results for the queries in Hive CLI:

   ```
   hive> SET hive.cli.print.header=true;
   ```

5. Optionally, create a `.hiverc` file in your home directory. Hive CLI will load this as an initialization script whenever you start the Hive CLI. You can set any customization properties such as enabling print headers (step 4) in this file. Other common usages of this file include switching to a Hive database and to register any libraries and custom UDFs (refer to the *Writing Hive User-defined Functions* recipe to learn more about UDFs) that you'll be using regularly.

## See also

Refer to `https://cwiki.apache.org/confluence/display/Hive/Configuration+Properties` for a complete list of configuration properties provided by Hive.

# Creating databases and tables using Hive CLI

This recipe walks you through the commands to create Hive databases and tables using the Hive CLI. Hive tables are used to define structure (schema) and other metadata information such as the location and storage format on datasets stored in HDFS. These table definitions enable the data processing and analysis using the Hive query language. As we discussed in the introduction, Hive follows a "schema on read" approach, where it imposes this structure only when reading and processing the data.

## Getting ready

For this recipe, you need a working Hive installation.

## How to do it...

This section depicts how to create a Hive table and how to perform simple queries on the Hive tables:

1. Start the Hive CLI by running the following command:

   ```
   $ hive
   ```

2. Execute the following command to create and use a Hive database for the Book-Crossing dataset mentioned in the introduction:

   ```
   hive> CREATE DATABASE bookcrossing;

   hive> USE bookcrossing;
   ```

3. Use the following command to view the details and the filesystem's location of the database:

   ```
   hive> describe database bookcrossing;

   OK

   bookcrossing    hdfs://……/user/hive/warehouse/bookcrossing.db
   ```

4. Let's create a table to map the user information data by running the following command in the Hive CLI. A table will be created inside the Book-Crossing database:

```
CREATE TABLE IF NOT EXISTS users
     (user_id INT,
     location STRING,
     age INT)
COMMENT 'Book Crossing users cleaned'
ROW FORMAT DELIMITED
FIELDS TERMINATED BY '\073'
STORED AS TEXTFILE;
```

5. Let's use the LOAD command to load the data to the table. The LOAD command copies the file to the Hive warehouse location in HDFS. Hive does not perform any data validation or data parsing at this step. Please note that the OVERWRITE clause in the Load command will overwrite and delete any old data that is already in the table:

```
hive> LOAD DATA LOCAL INPATH 'BX-Users-prepro.txt' OVERWRITE INTO
TABLE users;

Copying data from file:/home/tgunarathne/Downloads/BX-Users
-prepro.txt

Copying file: file:/home/tgunarathne/Downloads/BX-Users-prepro.txt

Loading data to table bookcrossing.users

Deleted /user/hive/warehouse/bookcrossing.db/users

Table bookcrossing.users stats: [num_partitions: 0, num_files: 1,
num_rows: 0, total_size: 10388093, raw_data_size: 0]

OK

Time taken: 1.018 seconds
```

6. Now, we can run a simple query to inspect the data in the created table. At this point, Hive parses the data using formats defined in the table definition and performs the processing specified by the query:

```
hive> SELECT * FROM users LIMIT 10;

OK

1   nyc, new york, usa   NULL

2   stockton, california, usa   18

3   moscow, yukon territory, russia   NULL

.........

10   albacete, wisconsin, spain   26

Time taken: 0.225 seconds, Fetched: 10 row(s)
```

7. Use the following command to view the columns of a Hive table:

```
hive> describe users;
OK
user_id                 int                     None
location                string                  None
age                     int                     None
Time taken: 0.317 seconds, Fetched: 3 row(s)
```

## How it works...

When we run Hive, we first define a table structure and load the data from a file into the Hive table. It is worth noting that the table definition must match the structure of the input data. Any type mismatches will result in NULL values, and any undefined columns will be truncated by Hive. The LOAD command copies the files into the Hive warehouse location without any changes, and these files will be managed by Hive. Table schema will be enforced on the data only when the data is read by Hive for processing:

```
CREATE TABLE IF NOT EXISTS users
    (user_id INT,
    location STRING,
    age INT)
COMMENT 'Book Crossing users cleaned'
ROW FORMAT DELIMITED
FIELDS TERMINATED BY '\073'
STORED AS TEXTFILE;
```

Please note that Hive table and column names are case insensitive. The preceding table will be created in the bookcrossing database as we issued the use bookcrossing command before issuing the create table command. Alternatively, you can also qualify the table name with the database name as bookcrossing.users. ROW FORMAT DELIMITED instructs Hive to use the native **SerDe** (Serializer and Deserializer classes, which are used by Hive to serialize and deserialize data) with delimited fields. In the dataset used for the preceding table, the fields are delimited using the ; character, which is specified using \073 because it is a reserved character in Hive. Finally, we instruct Hive that the data file is a text file. Refer to the *Utilizing different storage formats in Hive - storing table data using ORC files* recipe for more information on the different storage format options supported by Hive.

## There's more...

In this section, we explore Hive data types, Hive External tables, collection data types, and more about the describe command.

# Hive data types

A list of Hive data types that can be used to define the table columns can be found from `https://cwiki.apache.org/confluence/display/Hive/LanguageManual+Types`. These include simple data types such as TINYINT(1-byte signed integer), INT (4-byte signed integer), BIGINT (8-byte signed integer), DOUBLE (8-byte double precision floating point), TIMESTAMP, DATE, STRING, BOOLEAN, and several others.

Hive supports several complex collection data types such as arrays and maps as table column data types as well. Hive contains several built-in functions to manipulate the arrays and maps. One example is the `explode()` function, which outputs the items of an array or a map as separate rows. Refer to the *Using Hive built-in functions* recipe for more information on how to use Hive functions.

# Hive external tables

Hive external tables allow us to map a dataset in HDFS to a Hive table without letting Hive manage the dataset. Datasets for external tables will not get moved to the Hive default warehouse location.

Also, dropping an external table will not result in the deletion of the underlying dataset, as opposed to dropping a regular Hive table, where the dataset would get deleted. This is a useful feature when you want to prevent accidental deletion of data:

1. Copy the BX-Books-prepro.txt file to a directory in the HDFS:

   ```
   $ hdfs dfs -mkdir book-crossing
   $ hdfs dfs -mkdir book-crossing/books
   $ hdfs dfs -copyFromLocal BX-Books-prepro.txt book-crossing/books
   ```

2. Start the Hive CLI by running the following command and then use the Book-Crossing database:

   ```
   $ hive
   Hive> use bookcrossing;
   ```

3. Create an external table to map the book information data by running the following command in the Hive CLI:

   ```
   CREATE EXTERNAL TABLE IF NOT EXISTS books
       (isbn INT,
       title STRING,
       author STRING,
       year INT,
       publisher STRING,
       image_s STRING,
       image_m STRING,
   ```

```
    image_l STRING)
COMMENT 'Book crossing books list cleaned'
ROW FORMAT DELIMITED
FIELDS TERMINATED BY '\073'
STORED AS TEXTFILE
LOCATION '/user/<username>/book-crossing/books';
```

4. Use the following query to inspect the data of the newly created table:

```
hive> select * from books limit 10;

OK

195153448  Classical Mythology  Mark P. O. Morford  2002  Oxford
University Press  http://images.amazon.com/images/P/0195153448.01.
THUMBZZZ.jpg  http://images.amazon.com/images/P/0195153448.01.
MZZZZZZZ.jpg  http://images.amazon.com/images/P/0195153448.01.
LZZZZZZZ.jpg
```

5. Drop the table using the following command:

```
hive> drop table books;

OK

Time taken: 0.213 seconds
```

6. Inspect the data files in HDFS using the following command. Even though the table is dropped, the data files still exist:

```
$ hdfs dfs -ls book-crossing/books

Found 1 items

-rw-r--r--   1 tgunarathne supergroup   73402860 2014-06-19 18:49
/user/tgunarathne/book-crossing/books/BX-Books-prepro.txt
```

## Using the describe formatted command to inspect the metadata of Hive tables

You can use the `describe` command to inspect the basic metadata of the Hive tables. The `describe extended` command will print additional metadata information including the data location, input format, created time, and the like. The `describe formatted` command presents this metadata information in a more user-friendly manner:

```
hive> describe formatted users;
OK
```

| # col_name | data_type | comment |
| --- | --- | --- |
| user_id | int | None |
| location | string | None |

```
age                     int                      None


# Detailed Table Information
Database:               bookcrossing
Owner:                  tgunarathne
CreateTime:             Mon Jun 16 02:19:26 EDT 2014
LastAccessTime:         UNKNOWN
Protect Mode:           None
Retention:              0
Location:               hdfs://localhost:8020/user/hive/warehouse/
bookcrossing.db/users
Table Type:             MANAGED_TABLE
Table Parameters:
  comment               Book Crossing users cleaned
  numFiles              1
  numPartitions         0
  numRows               0
  rawDataSize           0
  totalSize             10388093
  transient_lastDdlTime 1402900035


# Storage Information
SerDe Library:          org.apache.hadoop.hive.serde2.lazy.LazySimpleSerDe
……
Time taken: 0.448 seconds, Fetched: 35 row(s)
```

# Simple SQL-style data querying using Apache Hive

We can query the datasets that have been mapped to Hive tables using HiveQL, which is similar to SQL. These queries can be simple data-exploration operations such as counts, orderby, and group by as well as complex joins, summarizations, and analytic operations. In this recipe, we'll explore simple data exploration Hive queries. The subsequent recipes in this chapter will present some of the advanced querying use cases.

## Getting ready

Install Hive and follow the earlier *Creating databases and tables using Hive CLI* recipe.

## How to do it...

This section demonstrates how to perform a simple SQL-style query using Hive.

1. Start Hive by issuing the following command:

   ```
   $ hive
   ```

2. Issue the following query in the Hive CLI to inspect the users aged between 18 and 34. Hive uses a MapReduce job in the background to perform this data-filtering operation:

   ```
   hive> SELECT user_id, location, age FROM users WHERE age>18 and
   age <34 limit 10;
   Total MapReduce jobs = 1
   Launching Job 1 out of 1
   ......
   10    albacete, wisconsin, spain   26
   13    barcelona, barcelona, spain   26
   ....
   Time taken: 34.485 seconds, Fetched: 10 row(s)
   ```

3. Issue the following query in the Hive CLI to count the total number of users that satisfy the above conditions (that is, whose ages are between 18 and 34). Hive converts this query to a MapReduce computation to calculate the result:

   ```
   hive> SELECT count(*) FROM users WHERE age>18 and age <34;
   Total MapReduce jobs = 1
   Launching Job 1 out of 1
   ...........
   2014-06-16 22:53:07,778 Stage-1 map = 100%,  reduce = 100%,
   ...........
   Job 0: Map: 1  Reduce: 1   Cumulative CPU: 5.09 sec   HDFS Read:
   10388330 HDFS Write: 6 SUCCESS
   Total MapReduce CPU Time Spent: 5 seconds 90 msec
   OK
   74339
   Time taken: 53.671 seconds, Fetched: 1 row(s)
   ```

4. The following query counts the number of users grouped by their age:

```
hive> SELECT  age, count(*) FROM users GROUP BY age;
Total MapReduce jobs = 1
.........
Job 0: Map: 1  Reduce: 1   Cumulative CPU: 3.8 sec   HDFS Read:
10388330 HDFS Write: 1099 SUCCESS
Total MapReduce CPU Time Spent: 3 seconds 800 msec
OK
....
10    84
11    121
12    192
13    885
14    1961
15    2376
```

5. The following query counts the number of users by their age and orders the result by the descending order of the number of users:

```
hive> SELECT  age, count(*) as c FROM users GROUP BY age ORDER BY
c DESC;
Total MapReduce jobs = 2
....
Job 0: Map: 1  Reduce: 1   Cumulative CPU: 5.8 sec   HDFS Read:
10388330 HDFS Write: 3443 SUCCESS
Job 1: Map: 1  Reduce: 1   Cumulative CPU: 2.15 sec   HDFS Read:
3804 HDFS Write: 1099 SUCCESS
Total MapReduce CPU Time Spent: 7 seconds 950 msec
OK
NULL  110885
24    5683
25    5614
26    5547
23    5450
27    5373
28    5346
29    5289
32    4778
```

## How it works...

You can use the `explain` command to view the execution plan of a Hive query. This is useful in identifying the bottlenecks of large-scale queries and in optimizing them. The following is the execution plan of one of the queries we used in the previous recipe. As you can see, this query resulted in a single MapReduce computation followed by a data output stage:

```
hive> EXPLAIN SELECT user_id, location, age FROM users WHERE age>18 and
age <34 limit 10;
OK
ABSTRACT SYNTAX TREE:

    ...


STAGE PLANS:
  Stage: Stage-1
    Map Reduce
      Alias -> Map Operator Tree:
        users
          TableScan
            alias: users
            Filter Operator
              predicate:
                  expr: ((age > 18) and (age < 34))
                  type: boolean
              Select Operator
                expressions:
                      expr: user_id
                      type: int
                      expr: location
                      type: string
                      expr: age
                      type: int
                outputColumnNames: _col0, _col1, _col2
                Limit
                  File Output Operator
                    compressed: false
                    GlobalTableId: 0
```

```
                        table:
                            input format: org.apache.hadoop.mapred.
TextInputFormat
                            output format: org.apache.hadoop.hive.ql.io.
HiveIgnoreKeyTextOutputFormat


    Stage: Stage-0
    Fetch Operator
        limit: 10
```

## There's more...

Hive provides several operators for the ordering of query results, with subtle differences and performance trade-offs:

- **ORDER BY**: This guarantees the global ordering of the data using a single reducer. However, for any non-trivial amount of result data, the use of a single reducer will significantly slow down your computation.

- **SORT BY**: This guarantees the local ordering of data that is output by each reduce task. However, the reduce tasks would contain overlapping data ranges.

- **CLUSTER BY**: This distributes the data to reduce tasks, avoiding any range overlaps, and each reduce task will output the data in a sorted order. This ensures the global ordering of data, even though the result will be stored in multiple files.

Refer to http://stackoverflow.com/questions/13715044/hive-cluster-by -vs-order-by-vs-sort-by for a more detailed explanation on the differences of the above mentioned operators.

## Using Apache Tez as the execution engine for Hive

Tez is a new execution framework built on top of YARN, which provides a lower-level API (directed acyclic graphs) than MapReduce. Tez is more flexible and powerful than MapReduce. Tez allows applications to improve performance by utilizing more expressive execution patterns than the MapReduce pattern. Hive supports the Tez execution engine as a substitute for the background MapReduce computations, where Hive would convert the Hive queries into Tez execution graphs, resulting in much-improved performance. You can perform the following procedures:

- You can instruct Hive to use Tez as the execution engine by setting the following hive property:

  ```
  hive> set hive.execution.engine=tez;
  ```

- You can switch back to MapReduce as the execution engine as follows:

  ```
  hive> set hive.execution.engine=mr;
  ```

## See also

Refer to `https://cwiki.apache.org/confluence/display/Hive/`
`LanguageManual+Select` for a list of clauses and features supported by the Hive
`select` statement.

# Creating and populating Hive tables and views using Hive query results

Hive allows us to save the output data of Hive queries by creating new Hive tables. We can
also insert the resultant data of a Hive query into another existing table as well.

## Getting ready

Install Hive and follow the *Creating databases and tables using Hive CLI* recipe.

## How to do it...

The following steps show you how to store the result of a Hive query into a new Hive table:

1. Issue the following query to save the output of the query of step 3 of the preceding
   recipe to a table named `tmp_users`:

   ```
   hive> CREATE TABLE tmp_users AS SELECT user_id, location, age FROM
   users WHERE age>18 and age <34;

   ...

   Table bookcrossing.tmp_users stats: [num_partitions: 0, num_files:
   1, num_rows: 0, total_size: 2778948, raw_data_size: 0]

   74339 Rows loaded to hdfs://localhost:8020/tmp/hive-root/
   hive_2014-07-08_02-57-18_301_5868823709587194356/-ext-10000
   ```

2. Inspect the data of the newly created table using the following command:

   ```
   hive> select * from tmp_users limit 10;
   OK
   10    albacete, wisconsin, spain   26
   13    barcelona, barcelona, spain   26
   18    rio de janeiro, rio de janeiro, brazil   25
   ```

3. Hive also allows us to insert the result of the Hive queries into an existing table as follows. Issue the following query to load the output data of the following query to the `tmp_users` Hive table:

```
hive> INSERT INTO TABLE tmp_users SELECT user_id, location, age
FROM users WHERE age>33 and age <51;

Total MapReduce jobs = 3

Launching Job 1 out of 3

.......

Loading data to table bookcrossing.tmp_users

Table bookcrossing.tmp_users stats: [num_partitions: 0, num_files:
2, num_rows: 0, total_size: 4717819, raw_data_size: 0]

52002 Rows loaded to tmp_users
```

4. You can also create a view in an existing table using a query as follows. The view can function as a regular table for query purposes, but the content of the view would get computed only on demand by Hive:

```
hive> CREATE VIEW tmp_users_view AS SELECT user_id, location, age
FROM users WHERE age>18 and age <34;
```

# Utilizing different storage formats in Hive - storing table data using ORC files

In addition to the simple text files, Hive also supports several other binary storage formats that can be used to store the underlying data of the tables. These include row-based storage formats such as Hadoop SequenceFiles and Avro files as well as column-based (columnar) storage formats such as ORC files and Parquet.

Columnar storage formats store the data column-by-column, where all the values of a column will be stored together as opposed to a row-by-row manner in row-based storages. For example, if we store the `users` table from our previous recipe in a columnar database, all the user IDs will be stored together and all the locations will be stored together. Columnar storages provide better data compression as it's easy to compress similar data of the same type that are stored together. Columnar storages also provide several performance improvements for Hive queries as well. Columnar storages allow the processing engine to skip the loading of data from columns that are not needed for a particular computation and also make it much faster to perform column-level analytical queries (for example, calculating the maximum age of the users).

In this recipe, we store the data from the `users` table of the *Creating databases and tables using Hive CLI* recipe into a Hive table stored in the ORC file format.

## Getting ready

Install Hive and follow the *Creating databases and tables using Hive CLI* recipe.

## How to do it...

The following steps show you how to create a Hive table stored using the ORC file format:

1.  Execute the following query in Hive CLI to create a user's table stored using the ORC file format:

    ```
    hive> USE bookcrossing;

    hive> CREATE TABLE IF NOT EXISTS users_orc
      (user_id INT,
      location STRING,
      age INT)
    COMMENT 'Book Crossing users table ORC format'
    STORED AS ORC;
    ```

2.  Execute the following command to insert the data into the newly created table. We have to populate the data using our earlier created table as we can't load text files directly to the ORC file or other storage format tables:

    ```
    hive> INSERT INTO TABLE users_orc
      SELECT *
      FROM users;
    ```

3.  Execute the following query to inspect the data in the `users_orc` table:

    ```
    Hive> select * from users_orc limit 10;
    ```

## How it works...

The `STORED AS ORC` phrase in the following command informs Hive that the data for this table will be stored using ORC files. You can use `STORED AS PARQUET` to store the table data using the Parquet format, and `STORED AS AVRO` to store the data using Avro files:

```
CREATE TABLE IF NOT EXISTS users_orc
  (user_id INT,
  location STRING,
  age INT)
STORED AS ORC;
```

# Using Hive built-in functions

Hive provides many built-in functions to aid us in the processing and querying of data. Some of the functionalities provided by these functions include string manipulation, date manipulation, type conversion, conditional operators, mathematical functions, and many more.

## Getting ready

This recipe assumes that the earlier recipe has been performed. Install Hive and follow the earlier recipe if you have not done already.

## How to do it...

This section demonstrates how to use the `parse_url` Hive function to parse the content of a URL:

1. Start Hive CLI by running the following command:

   ```
   $ hive
   ```

2. Issue the following command to obtain the `FILE` portion of the small image associated with each book:

   ```
   hive> select isbn, parse_url(image_s, 'FILE') from books limit 10;
   Total MapReduce jobs = 1

   …..

   OK
   0195153448   /images/P/0195153448.01.THUMBZZZ.jpg
   0002005018   /images/P/0002005018.01.THUMBZZZ.jpg
   0060973129   /images/P/0060973129.01.THUMBZZZ.jpg

   ……

   Time taken: 17.183 seconds, Fetched: 10 row(s)
   ```

## How it works...

The `parse_url` function gets invoked for each data record selected by the preceding query:

```
parse_url(string urlString, string partToExtract)
```

The `parse_url` function parses the URL given by the `urlString` parameter and supports HOST, PATH, QUERY, REF, PROTOCOL, AUTHORITY, FILE, and USERINFO as the `partToExtract` parameter.

## There's more...

You can issue the following command in the Hive CLI to see the list of functions supported by your Hive installation:

```
hive> show functions;
```

You can use the describe <function_name> and describe extended <function_name> commands in the Hive CLI as follows to access the help and usage for each of the functions. For example:

```
hive> describe function extended parse_url;
OK
parse_url(url, partToExtract[, key]) - extracts a part from a URL
Parts: HOST, PATH, QUERY, REF, PROTOCOL, AUTHORITY, FILE, USERINFO
key specifies which query to extract
Example:
  > SELECT parse_url('http://facebook.com/path/p1.php?query=1', 'HOST')
FROM src LIMIT 1;
  'facebook.com'
  > SELECT parse_url('http://facebook.com/path/p1.php?query=1', 'QUERY')
FROM src LIMIT 1;
  'query=1'
  > SELECT parse_url('http://facebook.com/path/p1.php?query=1', 'QUERY',
'query') FROM src LIMIT 1;
  '1'
```

## See also

Hive provides many different categories of functions, including mathematical, date manipulation, string manipulation, and many more. Refer to https://cwiki.apache.org/confluence/display/Hive/LanguageManual+UDF for a complete list of functions provided by Hive.

See the *Writing Hive User-defined Functions (UDF)* recipe for information on writing your own function to use with Hive queries.

# Hive batch mode - using a query file

In addition to the Hive interactive CLI, Hive also allows us to execute our queries in the batch mode, using a script file. In this recipe, we use a Hive script file to create books, users, and ratings tables of the Book-Crossing dataset and to load the data into the newly created tables.

## How to do it...

This section demonstrates how to create tables and load data using a Hive script file. Proceed with the following steps:

1. Extract the data package provided in the source repository of this chapter:

   ```
   $ tar -zxvf chapter6-bookcrossing-data.tar.gz
   ```

2. Locate the `create-book-crossing.hql` Hive query file in the Hive-scripts folder of the source repository for this chapter. Execute this Hive script file as follows by providing the location of the extracted data package for the `DATA_DIR` parameter. Please note that the execution of the following script file will overwrite any existing data in users, books, and ratings tables, if these exist beforehand, of the Book-Crossing database:

   ```
   $ hive \
       -hiveconf DATA_DIR=…/hcb-v2/chapter6/data/ \
       -f create-book-crossing.hql

   ……

   Copying data from file:……/hcb-v2/chapter6/data/BX-Books-Prepro.txt

   ……

   Table bookcrossing.books stats: [num_partitions: 0, num_files: 1,
   num_rows: 0, total_size: 73402860, raw_data_size: 0]
   OK

   ……

   OK

   Time taken: 0.685 seconds
   ```

3. Start the Hive CLI and issue the following commands to inspect the tables created by the preceding script:

   ```
   $ hive

   hive> use bookcrossing;

   ……

   hive> show tables;
   OK
   ```

```
books

ratings

users

Time taken: 0.416 seconds, Fetched: 3 row(s)

hive> select * from ratings limit 10;

OK

276725   034545104X   0

276726   0155061224   5

276727   0446520802   0
```

## How it works...

The `hive -f <filename>` option executes the HiveQL queries contained in the given file in a batch mode. With the latest Hive versions, you can even specify a file in HDFS as the script file for this command.

The create-book-crossing.hql script contains the commands to create the Book-Crossing database and to create and load data to users, books, and ratings tables:

```
CREATE DATABASE IF NOT EXISTS bookcrossing;
USE bookcrossing;

CREATE TABLE IF NOT EXISTS books
  (isbn STRING,
  title STRING,
  author STRING,
  year INT,
  publisher STRING,
  image_s STRING,
  image_m STRING,
  image_l STRING)
COMMENT 'Book crossing books list cleaned'
ROW FORMAT DELIMITED
FIELDS TERMINATED BY '\073'
STORED AS TEXTFILE;

LOAD DATA LOCAL INPATH '${hiveconf:DATA_DIR}/BX-Books-Prepro.txt'
OVERWRITE INTO TABLE books;
```

You can set properties and pass parameters to Hive script files using the `-hiveconf <property-name>=<property-value>` option when invoking the Hive command. You can refer to these properties inside the script using `${hiveconf:<property-name>}`. Such property usages inside the Hive queries will get substituted by the value of that property before the query executes. An example of this can be seen in the current recipe where we used the `DATA_DIR` property to pass the location of the data files to the Hive script. Inside the scripts, we used the value of this property using `${hiveconf:DATA_DIR}`.

The `-hiveconf` option can be used to set Hive configuration variables as well.

## There's more...

You can use the `hive -e '<query>'` option to run a batch mode Hive query directly from the command line. The following is an example of such a usage:

```
$ hive -e 'select * from bookcrossing.users limit 10'
```

## See also

Refer to `https://cwiki.apache.org/confluence/display/Hive/LanguageManual+Cli` for more information on the options supported by the Hive CLI.

# Performing a join with Hive

This recipe will guide you on how to use Hive to perform a join across two datasets. The first dataset is the book details dataset of the Book-Crossing database and the second dataset is the reviewer ratings for those books. This recipe will use Hive to find the authors with the most number of ratings of more than 3 stars.

## Getting ready

Follow the previous *Hive batch mode – using a query file* recipe.

## How to do it...

This section demonstrates how to perform a join using Hive. Proceed with the following steps:

1.  Start the Hive CLI and use the Book-Crossing database:

    ```
    $ hive

    hive > USE bookcrossing;
    ```

2. Create the books and book ratings tables by executing the `create-book-crossing.hql` Hive query file after referring to the previous *Hive batch mode commands using a query file* recipe. Use the following commands to verify the existence of those tables in the `Book-Crossing` database:

```
hive > SELECT * FROM books LIMIT 10;
....

hive > SELECT * FROM RATINGS LIMIT 10;
....
```

3. Now, we can join the two tables using Hive's SQL-like `join` command:

```
SELECT
    b.author AS author,
    count(*) AS count
FROM
    books b
JOIN
    ratings r
ON (b.isbn=r.isbn) and r.rating>3
GROUP BY b.author
ORDER BY count DESC
LIMIT 100;
```

4. If successful, it will print the following along with the results to the console:

```
Total MapReduce jobs = 4
...
2014-07-07 08:09:53  Starting to launch local task to process map
join;  maximum memory = 1013645312
...
Launching Job 2 out of 4
....
Launching Job 3 out of 4
...
2014-07-07 20:11:02,795 Stage-2 map = 100%,  reduce = 100%,
Cumulative CPU 8.18 sec
MapReduce Total cumulative CPU time: 8 seconds 180 msec
Ended Job = job_1404665955853_0013
Launching Job 4 out of 4
....
```

```
Total MapReduce CPU Time Spent: 21 seconds 360 msec
OK
Stephen King    4564
Nora Roberts    2909
John Grisham    2505
James Patterson  2334
J. K. Rowling   1742
...
Time taken: 116.424 seconds, Fetched: 100 row(s)
```

## How it works...

When executed, Hive first converts the join command into a set of MapReduce computations. These MapReduce computations will first load and parse both the datasets according to the given schema. Then, the data will be joined using the MapReduce computation according to the given join condition.

Hive supports inner joins as well as left, right, and full outer joins. Currently, Hive only supports equality-based conditions as join conditions. Hive is capable of performing several optimizations to optimize the performance of the joins based on the nature and size of the datasets.

## See also

▶ For more information, refer to `https://cwiki.apache.org/confluence/display/Hive/LanguageManual+Joins`.

▶ The *Joining two datasets using MapReduce* recipe of *Chapter 5*, *Analytics* shows how to implement a join operation using MapReduce.

# Creating partitioned Hive tables

This recipe will show how to use partitioned tables to store data in Hive. Partitioned tables allow us to store datasets partitioned by one or more data columns for efficient querying. The real data will reside in separate directories, where the names of the directories will form the values of the partition column. Partitioned tables can improve the performance of some queries by reducing the amount of data that Hive has to process by reading only select partitions when using an appropriate `where` predicate. A common example is to store transactional datasets (or other datasets with timestamps such as web logs) partitioned by the date. When the Hive table is partitioned by the date, we can query the data that belongs to a single day or a date range, reading only the data that belongs to those dates. In a non-partitioned table, this would result in a full table scan, reading all the data in that table, which can be very inefficient when you have terabytes of data mapped to a Hive table.

## Getting ready

This recipe assumes that the earlier *Hive batch mode - using a query file* recipe has been performed. Install Hive and follow the that recipe if you have not done already.

## How to do it...

This section demonstrates how to dynamically create a partitioned table in Hive. Proceed with the following steps:

1. Start Hive CLI.

2. Run the following commands to enable the dynamic partition creation in Hive:

   ```
   hive> set hive.exec.dynamic.partition=true;

   hive> set hive.exec.dynamic.partition.mode=nonstrict;

   hive> set hive.exec.max.dynamic.partitions.pernode=2000;
   ```

3. Execute the following query to create the new partitioned table using the results of the select statement. In this case, we partition the table using the published year of the books. Typically, years and dates serve as good partition columns for data that spans across time (for example, log data). When dynamically inserting data to a partitioned table, the partition column should be the last column in the insert statement:

   ```
   hive> INSERT INTO TABLE books_partitioned
       > partition (year)
       > SELECT
       >   isbn,
       >   title,
       >   author,
       >   publisher,
       >   image_s,
       >   image_m,
       >   image_l,
       >   year
       > FROM books;
   Total MapReduce jobs = 3
   Launching Job 1 out of 3
   ......

   Loading data to table bookcrossing.books_partitioned partition
   (year=null)
   ```

```
Loading partition {year=1927}

Loading partition {year=1941}

Loading partition {year=1984}

…….
```

```
Partition bookcrossing.books_partitioned{year=1982} stats: [num_
files: 1, num_rows: 0, total_size: 1067082, raw_data_size: 0]
```

…

4. Execute the following query. Due to the use of the `year` partition column, this query will only look at the data stored in the 1982 data partition of the Hive table. If not for the partitioning, this query would have required a MapReduce computation that processes the whole dataset:

```
hive> select * from books_partitioned where year=1982 limit 10;

OK

156047624  All the King's Men  Robert Penn Warren  Harvest
Books  http://images.amazon.com/images/P/0156047624.01.THUMBZZZ.
jpg  http://images.amazon.com/images/P/0156047624.01.MZZZZZZZ.jpg
http://images.amazon.com/images/P/0156047624.01.LZZZZZZZ.jpg  1982
```

5. Exit the Hive CLI and execute the following commands in the command prompt. You can see the partition directories created by Hive:

```
$ hdfs dfs -ls /user/hive/warehouse/bookcrossing.db/books_
partitioned

Found 116 items

drwxr-xr-x   - root hive          0 2014-07-08 20:24 /user/hive/
warehouse/bookcrossing.db/books_partitioned/year=0

drwxr-xr-x   - root hive          0 2014-07-08 20:24 /user/hive/
warehouse/bookcrossing.db/books_partitioned/year=1376

drwxr-xr-x   - root hive          0 2014-07-08 20:24 /user/hive/
warehouse/bookcrossing.db/books_partitioned/year=1378
```

….

# Writing Hive User-defined Functions (UDF)

As mentioned in the *Using Hive built-in functions* recipe, Hive supports many built-in functions for data manipulation and analysis. Hive also allows us to write our own customized functions to be used with the Hive queries. These functions are called user-defined functions, and this recipe will show you how to write a simple **User-defined Function** (**UDF**) for Hive. Hive UDFs allow us to extend the capabilities of Hive for our customized requirements, without having to resort to implementing Java MapReduce programs from scratch.

## Getting ready

This recipe assumes that the earlier recipe has been performed. Install Hive and follow the earlier recipe if you have not done already.

Make sure you have Apache Ant installed in your system.

## How to do it...

This section demonstrates how to implement a simple Hive UDF. Perform the following steps:

1. Use the Gradle build file in the source repository for this chapter to build the user-defined function JAR file:

   ```
   $ gradle build
   ```

2. Start Hive CLI:

   ```
   $ hive
   
   hive > USE bookcrossing;
   ```

3. Add the newly created JAR file to the environment using the full path of the JAR file created in step 1:

   ```
   hive> ADD JAR /home/../ hcb-c6-samples.jar;
   ```

4. Define the new UDF inside Hive using the following command:

   ```
   hive> CREATE TEMPORARY FUNCTION filename_from_url as 'chapter6.
   udf. ExtractFilenameFromURL';
   ```

5. Issue the following command to obtain the filename portion of the small image associated with each book using our newly defined UDF:

   ```
   hive> select isbn, filename_from_url(image_s, 'FILE') from books
   limit 10;
   ```

## How it works...

Hive UDFs should extend the UDF class of Hive and implement the `evaluate` method to perform the custom computation that you need to perform. The input and output parameters of the `evaluate` method needs to be provided using the appropriate Hadoop Writable type that corresponds to the Hive data type that you want to process and receive back from the UDF:

```
public class ExtractFilenameFromURL extends UDF {
  public Text evaluate(Text input) throws MalformedURLException {
    URL url = new URL(input.toString());
    Text fileNameText = new Text(FilenameUtils.getName(url.
getPath()));
```

```
        return fileNameText;
    }
}
```

We can use annotations like the following to add a description to the UDF. These would be emitted if you issue a `describe` command to this UDF from the Hive CLI:

```
@UDFType(deterministic = true)
@Description(
    name = "filename_from_url",
    value = "Extracts and return the filename part of a URL.",
    extended = "Extracts and return the filename part of a URL. "
        + "filename_from_url('http://test.org/temp/test.
jpg?key=value') returns 'test.jpg'."
)
```

# HCatalog – performing Java MapReduce computations on data mapped to Hive tables

HCatalog is a meta-data abstraction layer for files stored in HDFS and makes it easy for different components to process data stored in HDFS. HCatalog abstraction is based on tabular table model and augments structure, location, storage format and other meta-data information for the data sets stored in HDFS. With HCatalog, we can use data processing tools such as Pig, Java MapReduce and others read and write data to Hive tables without worrying about the structure, storage format or the storage location of the data. HCatalog is very useful when you want to execute a Java MapReduce job or a Pig script on a data set that is stored in Hive using a binary data format such as ORCFiles. The topology can be seen as follows:

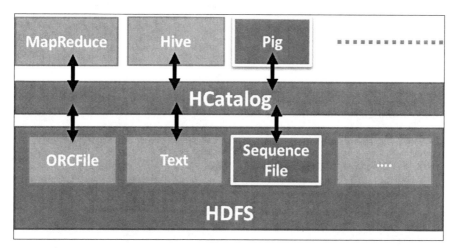

HCatalog achieves this capability by providing an interface to the Hive MetaStore enabling the other applications to utilize the Hive table metadata information. We can query the table information in HCatalog using the HCatalog Command Line Interface (CLI). HCatalog CLI is based on Hive CLI and supports Hive **Data Definition Language** (**DDL**) statements except for statements that require running a MapReduce query. HCatalog also exposes a REST API called WebHCat.

In this recipe we'll be looking at using Java MapReduce computations on top of the data stored in Hive tables by utilizing the meta-data available from HCatalog. HCatalog provides HCatInputFormat class to retrieve data from the Hive tables.

## Getting ready

Make sure HCatalog is installed with Hive in your system.

## How to do it...

This section demonstrates how to process Hive table data using MapReduce computations. Perform the following steps:

1.  Follow the Hive batch mode commands using a query file recipe of this chapter to create and populate the `bookcrossing.user` Hive table that we'll be using in this recipe.

2.  Compile the sample source code for this chapter by running the following `gradle` command from the `chapter6` folder of the source repository.

    ```
    $ gradle clean build uberjar
    ```

3.  Run the MapReduce job using the following command. The first parameter is the database name, second parameter is the input table name and the third parameter is the output path. This job counts the number of users aged between 18 and 34 grouped by the year:

    ```
    $ hadoop jar build/libs/hcb-c6-samples-uber.jar \
    chapter7.hcat.HCatReadMapReduce \
    bookcrossing users hcat_read_out
    ```

4.  Inspect the results of this computation by running the following command:

    ```
    $ hdfs dfs -cat hcat-read-out/part*
    ```

## How it works...

You can find the source code for this recipe from `chapter6/hcat/ HCatReadMapReduce. java` file in the source folder of this chapter.

Following lines in the `run()` function specify the `HCatalogInputFormat` as the `InputFormat` for the computation and configures it with the input database name and table name.

```
// Set HCatalog as the InputFormat
job.setInputFormatClass(HCatInputFormat.class);
HCatInputFormat.setInput(job, dbName, tableName);
```

The `map()` function receives the records from the Hive table as `HCatRecord` values, while the `map()` input key does not contain any meaningful data. HCatRecord contains the data fields parsed according to the column structure of the Hive table and we can extract the fields from the `HCatRecord` as follows in the `map` function:

```
public void map( WritableComparable key,HCatRecord value,…)
        throws IOException, InterruptedException {
  HCatSchema schema = HCatBaseInputFormat.
        getTableSchema(context.getConfiguration());
  // to avoid the "null" values in the age field
  Object ageObject = value.get("age", schema);
  if (ageObject instanceof Integer) {
    int age = ((Integer) ageObject).intValue();
    // emit age and one for count
    context.write(new IntWritable(age), ONE);
    }
     }
    }
```

HCatalog jars, Hive jars and their dependencies are needed in the Hadoop Classpath to execute the HCatalog MapReduce programs. We also need to supply these jars to Map and Reduce tasks by specifying the dependency libraries using the `libjars` parameter at the command line when invoking the Hadoop JAR command. An alternative to solve both the Hadoop Classpath and the `libjars` requirements is to package all the dependency jars in to a single fat-jar and use it to submit the MapReduce program.

In this sample, we use the second approach and create a fat-jar (`hcb-c6-samples-uber.jar`) using the Gradle build as follows:

```
task uberjar(type: Jar) {
  archiveName = "hcb-c6-samples-uber.jar"
  from files(sourceSets.main.output.classesDir)
  from {configurations.compile.collect {zipTree(it)}} {
      exclude "META-INF/*.SF"
      exclude "META-INF/*.DSA"
      exclude "META-INF/*.RSA"
  }
}
```

# HCatalog – writing data to Hive tables from Java MapReduce computations

HCatalog also allows us to write data to Hive tables from Java MapReduce computations using the `HCatOutputFormat`. In this recipe, we'll be looking at how to write data to a Hive table using a Java MapReduce computation. This recipe extends the computation of the previous *HCatalog – performing Java MapReduce computations on data mapped to Hive tables* recipe by adding table write capability.

## Getting ready

Make sure HCatalog is installed with Hive in your system.

## How to do it...

This section demonstrates how to write data to a Hive table using a MapReduce computation. Perform the following steps:

1. Follow the *Hive batch mode – using a query file* recipe of this chapter to create and populate the user Hive table that we'll be using in this recipe.

2. Compile the sample source code for this chapter by running the following `gradle` command from the `chapter6` folder of the source repository:

   `$ gradle clean build uberjar`

3. Use Hive CLI to create a hive table to store the results of the computation.

   `hive> create table hcat_out(age int, count int);`

4. Run the MapReduce job using the following command. The first parameter is the database name, second parameter is the input table name and the third parameter is the output table name. This job counts the number of users aged between 18 and 34 grouped by the year and writes the results to the hcat_out table we created in step 3:

```
$ hadoop jar hcb-c6-samples-uber.jar \
chapter6.hcat.HCatWriteMapReduce \
bookcrossing users hcat_out
```

5. Read the results by running the following command in the hive CLI:

```
hive> select * from bookcrossing.hcat_out limit 10;
OK
hcat_out.age      hcat_out.count
19      3941
20      4047
21      4426
22      4709
23      5450
24      5683
```

## How it works...

You can find the source for the recipe from `chapter6/src/chapter6/hcat/HCatWriteMapReduce.java` file.

In addition to the configurations we discussed in the previous *HCatalog – performing Java MapReduce computations on data mapped to Hive tables* recipe, we specify `HCatalogOutputFormat` as the `OutputFormat` for the computation in the `run()` function as follows. We also configure the output database and table name:

```
job.setOutputFormatClass(HCatOutputFormat.class);

HCatOutputFormat.setOutput(job,
    OutputJobInfo.create(dbName, outTableName, null));
```

We have to use the `DefaultHCatRecord` as the job output value when writing data to a Hive table:

```
job.setOutputKeyClass(WritableComparable.class);
job.setOutputValueClass(DefaultHCatRecord.class);
```

We set the schema for the output table as follows:

```
HCatSchema schema = HCatOutputFormat.getTableSchema(job
                          .getConfiguration());
HCatOutputFormat.setSchema(job, schema);
```

The `reduce()` function outputs the data as `HCatRecord` values. `HCatOutputFormat` ignores any output keys:

```
public void reduce(IntWritable key, Iterable<IntWritable> values,
        Context context) ... {
  if (key.get() < 34 & key.get() > 18) {
     int count = 0;
     for (IntWritable val : values) {
   count += val.get();
     }

      HCatRecord record = new DefaultHCatRecord(2);
      record.set(0, key.get());
      record.set(1, count);
      context.write(null, record);
   }
}
```

# 7
# Hadoop Ecosystem II – Pig, HBase, Mahout, and Sqoop

In this chapter, we will cover the following topics:

- ▶ Getting started with Apache Pig
- ▶ Joining two datasets using Pig
- ▶ Accessing a Hive table data in Pig using HCatalog
- ▶ Getting started with Apache HBase
- ▶ Data random access using Java client APIs
- ▶ Running MapReduce jobs on HBase
- ▶ Using Hive to insert data into HBase tables
- ▶ Getting started with Apache Mahout
- ▶ Running K-means with Mahout
- ▶ Importing data to HDFS from a relational database using Apache Sqoop
- ▶ Exporting data from HDFS to a relational database using Apache Sqoop

## Introduction

Hadoop ecosystem has a family of projects that are either built on top of Hadoop or work very closely with Hadoop. These projects have given rise to an ecosystem that focuses on large-scale data processing, and often users can use several of these projects in combination to solve their big data problems.

This chapter introduces several key projects in the Hadoop ecosystem and shows how to get started with each of these projects.

We will focus on the following four projects:

- **Pig**: A dataflow-style data processing language for large-scale processing of data stored in HDFS

- **HBase**: A NoSQL-style highly scalable data store, which provides low latency, random accessible and highly scalable data storage on top of HDFS

- **Mahout**: A toolkit of machine-learning and data-mining tools

- **Sqoop**: A data movement tool for efficient bulk data transfer between Apache Hadoop ecosystem and ralational databases

Some of the HBase and Mahout recipes of this chapter are based on the *Chapter 5, Hadoop Ecosystem* chapter of the previous edition of this book, *Hadoop MapReduce Cookbook*. Those recipes were originally authored by Srinath Perera.

**Sample code**

The sample code and data files for this book are available in GitHub at `https://github.com/thilg/hcb-v2`. The `chapter7` folder of the code repository contains the sample code for this chapter.

Sample codes can be compiled and built by issuing the gradle build command in the `chapter7` folder of the code repository. Project files for Eclipse IDE and IntelliJ IDEA IDE can be generated by running the `gradle eclipse` and `gradle idea` commands respectively in the main folder of the code repository.

Some of the recipes of this chapter use the Book Crossing dataset as the sample data. This dataset is compiled by Cai-Nicolas Ziegler and comprises a list of books, list of users, and a list of ratings. The `chapter6` folder of the source repository contains a cleaned sample of this dataset. You can obtain the full dataset from `http://www2.informatik.uni-freiburg.de/~cziegler/BX/`.

# Getting started with Apache Pig

Apache Pig is a high-level language framework for Hadoop that makes it easy to analyze very large datasets stored in HDFS without having to implement complex Java MapReduce applications. The language of Pig is called Pig Latin, which is a data flow language. While the goal of both Pig and Hive frameworks is similar, the language layers of these two frameworks complement each other by providing a procedural language and a declarative language, respectively.

Pig converts Pig Latin queries in to a series of one or more MapReduce jobs in the background.

In order to install Pig, we recommend you use one of the freely available commercial Hadoop distributions as described in *Chapter 1, Getting Started with Hadoop v2*. Another alternative is to use Apache Bigtop to install Pig. Refer to the Bigtop-related recipe in *Chapter 1, Getting Started with Hadoop v2* for steps on installing Pig using the Apache Bigtop distribution.

In case you don't have a working Pig and Hadoop installation, the following steps show you how to install Pig with MapReduce local mode using the local file system as the data storage. This is recommended only for learning and testing purposes.

Download and extract the latest version of Pig from `http://pig.apache.org/releases.html`. Add the `bin` directory of the extracted folder to your PATH environment variable as follows:

```
$ export PATH=pig-0.13.0/bin:$PATH
```

Use the `pig` command with the `local` flag to start the Grunt shell, as follows:

```
$ pig -x local
grunt>
```

This recipe demonstrates how to use Pig queries to process data in HDFS. We will use the BookCrossing dataset for this recipe. This recipe will use Pig to process the BookCrossing user dataset and select a list of users who are aged between 18 and 34, ordered by their age.

## Getting ready

This recipe requires a working Pig installation integrated with a Hadoop YARN cluster. You can run these samples using the Pig local mode as well. However, in such cases you'll have to use the local file system instead of HDFS to load the data.

## How to do it...

This section describes how to use Pig Latin queries to find users aged between 18 and 34 sorted by the age from the BookCrossing user dataset. Proceed with the following steps:

1.  Copy and extract the BookCrossing sample dataset (`chapter6-bookcrossing -data.tar.gz`) from the `chapter6` folder of the code repository.

2.  Create a directory in HDFS and copy the BookCrossing user dataset into that directory, as follows:

    ```
    $ hdfs dfs -mkdir book-crossing

    $ hdfs dfs -copyFromLocal \
    chapter6/data/BX-Users-Prepro.txt book-crossing
    ```

3.  Start the Pig Grunt shell and issue the following Pig commands:

    ```
    $ pig

    grunt> A = LOAD 'book-crossing/BX-Users-Prepro.txt' USING
    PigStorage(';')  AS (userid:int, location:chararray, age:int);

    grunt> B = FILTER A BY age > 18 AND age < 34 ;

    grunt> C = ORDER B BY age;
    ```

4.  Print the output of the processing flow by using the DUMP operator in the same grunt shell. The queries we issued in step 3 get executed only after we issue the following command (or any other data output command). You should notice a series of MapReduce jobs after issuing the following two commands:

    ```
    grunt> D = LIMIT C 10;

    grunt> DUMP D;
    ```

    The output of the preceding command is as follows:

    ```
    2015-01-28 07:06:57,745 [main] INFO  org.apache.hadoop.m
    2015-01-28 07:06:57,745 [main] INFO  org.apache.pig.back
    (11882,brockville, ontario, canada,19)
    (88601,chilliwack, british columbia, canada,19)
    (191586,richmond, virginia, usa,19)
    (227883,florissant, missouri, usa,19)
    (238161,glasgow, lanarkshire, united kingdom,19)
    (110435,crown point, indiana, usa,19)
    (7441,north vancouver, british columbia, canada,19)
    (110444,são paulo, são paulo, brazil,19)
    (221868,montague, michigan, usa,19)
    (249221,concord, california, usa,19)
    grunt> █
    ```

5. You can also use the ILLUSTRATE operator to test your queries. The Illustrate operator retrieves a small sample of data from your input data and runs your queries on that data, giving faster turnaround times to review and test the Pig queries:

```
grunt> ILLUSTRATE B;
```

The output of the preceding command is as follows:

```
2015-01-28 10:16:52,333 [main] INFO  org.apache.pig.backend.hadoop.executionengine.mapReduceLayer
essed per job phase (AliasName[line,offset]): M: A[1,4],A[-1,-1],B[2,4] C:  R:
-----------------------------------------------------------------------------
| A     | userid:int    | location:chararray                     | age:int   |
-----------------------------------------------------------------------------
|       | 2749          | bridge lake, british columbia, canada  | 18        |
|       | 3345          | bishopscastle, england, united kingdom | 24        |
|       | 4436          | neath, wales, united kingdom           | 43        |
-----------------------------------------------------------------------------

-----------------------------------------------------------------------------
| B     | userid:int    | location:chararray                     | age:int   |
-----------------------------------------------------------------------------
|       | 3345          | bishopscastle, england, united kingdom | 24        |
-----------------------------------------------------------------------------

grunt>
```

## How it works...

When we issue Pig queries, Pig internally converts them to a set of MapReduce jobs and executes them in the Hadoop cluster to obtain the desired result. For almost all the data queries, Pig queries are much easier to write and manage than MapReduce applications.

The following line instructs Pig to load the data file in to a relation named A. We can provide either a single file or a directory to the load command. Using PigStorage(';') instructs Pig to load the data using the default load function with ; as the separator. When the MapReduce job is executed, Pig's load function parses the input data and assigns it to the fields of the schema described in the AS clause. Any data point that doesn't fit in to the given schema would result in an error or a NULL value at the time of execution:

```
grunt> A = LOAD 'book-crossing/BX-Users-Prepro.txt' USING
PigStorage(';')  AS (userid:int, location:chararray, age:int);
```

The FILTER operator selects data from the relation based on a given condition. In the following line of code, we select the data points, where the age of the user is between 18 and 34:

```
grunt> B = FILTER A BY age > 18 AND age < 34;
```

The ORDER BY operator sorts the data in a relation based on one or more data fields. In the following query, we sort the relation B by the age of the user:

```
grunt> C = ORDER B BY age;
```

The `LIMIT` operator limits the number of data points (tuples) in a relation using the given number. In the following query, we limit relation C to only 10 tuples. The following step makes it easy to inspect the data using the DUMP operator:

```
grunt> D = LIMIT C 10;
```

## There's more...

Pig Latin also contains a large set of built-in functions providing functionalities in the areas of math, string processing, data time processing, basic statistics, data loading and storing, and several others. A list of Pig's built-in functions can be found from `www.pig.apache.org/docs/r0.13.0/func.html`. You can also implement Pig `User Defined Functions` to perform any custom processing that you require.

## See also

- Refer to `www.pig.apache.org/docs/r0.13.0/basic.html` for documentation on Pig Latin data types, operators, and other basics
- Refer to `www.pig.apache.org/docs/r0.13.0/test.html` for documentation on Pig Latin testing operators such as `ILLUSTRATE`, `DUMP`, and `DESCRIBE`

# Joining two datasets using Pig

This recipe explains how to join two datasets using Pig. We will use the BookCrossing dataset for this recipe. This recipe will use Pig to join the Books dataset with the Book-Ratings dataset and find the distribution of high ratings (with rating>3) with respect to authors.

## How to do it...

This section describes how to use a Pig Latin script to find author's review rating distribution by joining the Books dataset with the Ratings dataset:

1. Extract the BookCrossing sample dataset (`chapter6-bookcrossing-data.tar.gz`) from the `chapter6` folder of the code repository.

2. Create a directory in HDFS and copy the BookCrossing Books dataset and the Book-Ratings dataset to that directory, as follows:

```
$ hdfs dfs -mkdir book-crossing

$ hdfs dfs -copyFromLocal \
chapter6/data/BX-Books-Prepro.txt book-crossing

$ hdfs dfs -copyFromLocal \
BX-Book-Ratings-Prepro.txt book-crossing
```

3. Review the `chapter7/pig-scripts/book-ratings-join.pig` script.

4. Execute the preceding Pig Latin script using the following command:

```
$ pig -f pig-scripts/book-ratings-join.pig
```

The output of the preceding command is as follows:

```
2015-01-28 10:47:58,524 [main] INFO  org.apache.hadoop.mapreduce.lib.
2015-01-28 10:47:58,524 [main] INFO  org.apache.pig.backend.hadoop.exe
(1,N)
(3,X)
(1,Ai)
(2,Bh)
(6,Na)
(1,Ty)
(1,ky)
(2,tk)
(3,AAA)
(1,AHD)
(86,Avi)
(4,Bbc)
(1,Bez)
(1,Bhg)
```

## How it works...

The following Pig commands load the data to the Books and BookRatings relations. As described in the earlier recipe, `PigStorage(';')` instructs Pig to use '; ' as the field separator:

```
Books = LOAD 'book-crossing/BX-Books-Prepro.txt'
USING PigStorage(';')  AS (
    isbn:chararray,
    title:chararray,
    author:chararray,
    year:int,
    publisher:chararray,
    image_s:chararray,
    image_m:chararray,
    image_l:chararray);
Ratings = LOAD 'book-crossing/BX-Book-Ratings-Prepro.txt'
  USING PigStorage(';')  AS (
    userid:int,
    isbn:chararray,
    ratings:int);
```

We select only the reviews with good ratings using the following `FILTER` operation:

```
GoodRatings = FILTER R BY ratings > 3;
```

Then, we join the Books and GoodRatings relations using ISBN as the common criteria. This is an inner equi join and produces a Cartesian product of all the records filtered by the join criteria. In other words, the resultant relation contains a record for each matching book and a book rating (number of matching books X number of good ratings):

```
J = JOIN Books BY isbn, GoodRatings by isbn;
```

The following statement groups the join result by the author. Each group now contains all the records belonging to an author. Assuming we have a matching book for each good rating, the number of records in a group would be the number of good reviews the author of that group has received.

```
JA = GROUP J BY author;
```

The following statement counts the number of records in each group of relation JA and output the author name and the count of good reviews for books written by that author:

```
JB = FORACH JA GENERATE group, COUNT(J);
OA = LIMIT JB 100;
DUMP OA;
```

You can manually issue the preceding commands in the Pig Grunt shell to gain a more detailed understanding of the data flow. While doing so, you can use `LIMIT` and `DUMP` operators to understand the result of each step.

## There's more...

Pig supports outer joins as well. However, currently Pig only supports equi joins, where the join condition has to be based on equality.

# Accessing a Hive table data in Pig using HCatalog

There can be scenarios where we want to access the same dataset from both Hive and Pig. There can also be scenarios where we want to process the results of a Hive query that's mapped to a Hive table using Pig. In such cases, we can take advantage of the HCatalog integration in Pig to access HCatalog managed Hive tables from Pig without worrying about the data definition, data storage format, or the storage location.

Follow the *Hive batch mode - using a query file* recipe from *Chapter 6, Hadoop Ecosystem – Apache Hive* to create the Hive table that we'll be using in this recipe.

## How to do it...

This section demonstrates how to access a Hive table from Pig. Proceed with the following steps:

1. Start the Pig's Grunt shell with the `-useHCatalog` flag, as follows. This will load the HCatalog JARs that are necessary to access HCatalog managed tables in Hive:

   ```
   $ pig -useHCatalog
   ```

2. Use the following command in the Grunt shell to load the `users` table from the `bookcrossing` Hive database into a Pig relation named `users`. HCatLoader facilitates the reading of data from HCatalog managed tables:

   ```
   grunt> users = LOAD 'bookcrossing.users' USING org.apache.hive.
   hcatalog.pig.HCatLoader();
   ```

3. Use the `describe` operator as follows to inspect the schema of the `users` relation:

   ```
   grunt> DESCRIBE users;

   users: {user_id: int,location: chararray,age: int}
   ```

4. Inspect the data of the `users` relation by issuing the following command in the Pig Grunt shell. The relations loaded through Hive can be used similarly to any other relation in Pig:

   ```
   grunt> ILLUSTRATE users;
   ```

   The output of the preceding command is as follows:

   ```
   2015-01-28 11:30:48,937 [main] INFO  org.apache.pig.backend.hadoop.executione
   essed per job phase (AliasName[line,offset]): M: users[1,8] C:  R:
   -------------------------------------------------------------------------------
   | users     | user_id:int  | location:chararray            | age:int  |
   -------------------------------------------------------------------------------
   |           | 4715         | villaalemana, valparaiso, chile | 20      |
   -------------------------------------------------------------------------------
   grunt>
   ```

## There's more...

You can also store data in Hive tables from Pig using the HCatStorer interface to write data to HCatalog managed tables, as follows:

```
grunt> STORE r INTO 'database.table'
        USING org.apache.hcatalog.pig.HCatStorer();
```

## See also

The *HCatalog – performing Java MapReduce computations on data mapped to Hive tables* and *HCatalog – writing data to Hive tables from Java MapReduce computations* recipes of *Chapter 6, Hadoop Ecosystem – Apache Hive*.

# Getting started with Apache HBase

HBase is a highly scalable distributed NoSQL data store that supports columnar-style data storage. HBase is modeled after Google's Bigtable. HBase uses HDFS for data storage and allows random access of data, which is not possible in HDFS.

The HBase table data model can be visualized as a very large multi-dimensional sorted map. HBase tables consist of rows, each of which has a unique Row Key, followed by a list of columns. Each row can have any number of columns and doesn't have to adhere to a fixed schema. Each data cell (column in a particular row) can have multiple values based on timestamps, resulting in a three-dimensional table (row, column, timestamp). HBase stores all the rows and columns in a sorted order making it possible to randomly access the data.

Although the data model has some similarities with the relational data model, unlike relational tables, different rows in the HBase data model may have different columns. For instance, the second row may contain completely different name-value pairs from the first one. HBase also doesn't support transactions or atomicity across the rows. You can find more details about this data model from the Google's Bigtable paper, `http://research.google.com/archive/bigtable.html`.

HBase supports the storage of very large datasets and provides low-latency high-throughput reads and writes. HBase powers some of the very demanding real-time data processing systems such as online advertisement agencies; it has powered Facebook Messenger as well. The data stored in HBase can also be processed using MapReduce.

HBase cluster architecture consists of one or more master nodes and a set of region servers. HBase tables are horizontally split into regions, which are served and managed by region servers. Regions are further broken down vertically by column families and saved in HDFS as files. Column families are a logical grouping of columns in a table, which results in physical grouping of columns at the storage layer.

Obtaining the maximum performance out of HBase requires careful designing of tables, taking its distributed nature in to consideration. RowKeys play an important role in the performance as the region distribution and any querying is based on RowKeys. Recipes in this book do not focus on such optimizations.

In order to install HBase, we recommend that you use one of the freely available commercial Hadoop distributions as described in *Chapter 1, Getting Started with Hadoop v2*. Another alternative is to use an HBase cluster on the Amazon cloud environment as described in *Chapter 2, Cloud Deployments – Using Hadoop YARN on Cloud Environments*.

## Getting ready

This recipe requires an Apache HBase installation integrated with a Hadoop YARN cluster. Make sure to start all the configured HBase Master and RegionServer processes before we begin.

## How to do it...

This section demonstrates how to get started with Apache HBase. We are going to create a simple HBase table and insert a row of data to that table using the HBase shell. Proceed with the following steps:

1. Start the HBase shell by executing the following command:

   ```
   $ hbase shell
   ......
   hbase(main):001:0>
   ```

2. Issue the following command in the HBase shell to check the version:

   ```
   hbase(main):002:0> version
   0.98.4.2.2.0.0-2041-hadoop2,
   r18e3e58ae6ca5ef5e9c60e3129a1089a8656f91d, Wed Nov 19 15:10:28 EST
   2014
   ```

3. Create an HBase table named `test` table. List all the tables to verify the creation of the `test` table, as follows:

   ```
   hbase(main):003:0> create 'test', 'cf'
   0 row(s) in 0.4210 seconds
   => Hbase::Table - test

   hbase(main):004:0> list
   TABLE
   SYSTEM.CATALOG
   SYSTEM.SEQUENCE
   ```

```
SYSTEM.STATS
test
4 row(s) in 0.0360 seconds
```

4. Now, insert a row to the `test` table using the HBase `put` command as follows. Use `row1` as the RowKey, `cf:a` as the column name and 10 as the value

```
hbase(main):005:0> put 'test', 'row1', 'cf:a', '10'
0 row(s) in 0.0080 seconds
```

5. Scan the `test` table using the following command, which prints all the data in the table:

```
hbase(main):006:0> scan 'test'
ROW          COLUMN+CELL
row1column=cf:a, timestamp=1338485017447, value=10
1 row(s) in 0.0320 seconds
```

6. Retrieve the value from the table using the following command by giving `test` as the table name and `row1` as RowKey:

```
hbase(main):007:0> get 'test', 'row1'
COLUMN       CELL
cf:atimestamp=1338485017447, value=10
1 row(s) in 0.0130 seconds
```

7. Disable and drop the test table using the `disable` and `drop` commands, as follows:

```
hbase(main):014:0> disable 'test'
0 row(s) in 11.3290 seconds

hbase(main):015:0> drop 'test'
0 row(s) in 0.4500 seconds
```

## There's more...

In addition to the next several recipes in this chapter, the following recipes in this book also use HBase and provide more use cases for HBase:

▶ The *Loading large datasets to an Apache HBase data store - importtsv and bulkload* recipe of *Chapter 10, Mass Text Data Processing*

▶ The *Creating TF and TF-IDF vectors for the text data* recipe of *Chapter 10, Mass Text Data Processing*

▶ The *Generating the in-links graph for crawled web pages* recipe of *Chapter 8, Searching and Indexing*

▶ The *Deploying an Apache HBase cluster on Amazon EC2 using EMR* recipe of *Chapter 2, Cloud Deployments – Using Hadoop YARN on Cloud Environments*

## See also

▶ Extensive documentation on HBase is available at `http://hbase.apache.org/book.html`.

# Data random access using Java client APIs

The previous recipe introduced the command-line interface for HBase. This recipe demonstrates how we can use the Java API to interact with HBase.

## Getting ready

This recipe requires an Apache HBase installation integrated with a Hadoop YARN cluster. Make sure to start all the configured HBase Master and RegionServer processes before we begin.

## How to do it...

The following step executes an HBase Java client to store and retrieve data from an HBase table.

Run the `HBaseClient` Java program by running the following command from the `chapter 7` folder of the sample source repository:

```
$ gradle execute HBaseClient
```

## How it works...

The source code for the preceding Java program is available in the `chapter7/src/chapter7/hbase/HBaseClient.java` file in the source repository. The following code creates an HBase configuration object and then creates a connection to the `test` HBase table. This step obtains the HBase hostnames and ports using ZooKeeper. In high throughput production scenarios, it's recommended to connect to HBase tables using `HConnection` instances.

```
Configuration conf = HBaseConfiguration.create();
HTable table = new HTable(conf, "test");
```

The following commands will add a data row to the HBase table:

```
Put put = new Put("row1".getBytes());
put.add("cf".getBytes(), "b".getBytes(), "val2".getBytes());
table.put(put);
```

Search for data by performing a scan, as follows:

```
Scan s = new Scan();
s.addFamily(Bytes.toBytes("cf"));
ResultScanner results = table.getScanner(s);
```

# Running MapReduce jobs on HBase

This recipe explains how to run a MapReduce job that reads and writes data directly to and from HBase storage.

HBase provides abstract mapper and reducer implementations that users can extend to read and write directly from HBase. This recipe explains how to write a sample MapReduce application using these mappers and reducers.

We will use the World Bank's **Human Development Report** (**HDR**) data, by country, which shows **Gross National Income** (**GNI**) per capita of each country. The dataset can be found at `http://hdr.undp.org/en/statistics/data/`. A sample of this dataset is available in the `chapter7/resources/hdi-data.csv` file in the sample source code repository. Using MapReduce, we will calculate average value of GNI per capita, by country.

## Getting ready

This recipe requires an Apache HBase installation integrated with a Hadoop YARN cluster. Make sure to start all the configured HBase Master and RegionServer processes before we begin.

## How to do it...

This section demonstrates how to run a MapReduce job on data stored in HBase. Proceed with the following steps:

1. Execute the `gradle build` command from the `chapter7` folder of the source repository to compile the source code, as follows:

   ```
   $ gradle build
   ```

2. Run the following command from the `chapter7` folder to upload the sample data to HBase. This command uses the `chapter7/src/chapter7/hbase/HDIDataUploader` to upload the data:

```
$ gradle executeHDIDataUpload
```

3. Run the MapReduce job by running the following command from `HADOOP_HOME`:

```
$ hadoop jar hcb-c7-samples.jar \
    chapter7.hbase.AverageGINByCountryCalcualtor
```

4. View the results in HBase by running the following command from the HBase shell:

```
$ hbase shell
hbase(main):009:0> scan  'HDIResult'
```

## How it works...

You can find the Java HBase MapReduce sample in `chapter7/src/chapter7/hbase/AverageGINByCountryCalcualtor.java`. Since we are going to use HBase to read the input as well as to write the output, we use the HBase `TableMapper` and `TableReducer` helper classes to implement our MapReduce application. We configure the `TableMapper` and the `TableReducer` using the utility methods given in the `TableMapReduceUtil` class. The `Scan` object is used to specify the criteria to be used by the mapper when reading the input data from the HBase data store.

# Using Hive to insert data into HBase tables

Hive-HBase integration gives us the ability to query HBase tables using the **Hive Query Language** (**HQL**). Hive-HBase integration supports mapping of existing HBase tables to Hive tables as well as the creation of new HBase tables using HQL. Both reading data from HBase tables and inserting data into HBase tables are supported through HQL, including performing joins between Hive-mapped HBase tables and traditional Hive tables.

The following recipe uses HQL to create an HBase table to store the `books` table of the `bookcrossing` dataset and populate that table using sample data.

## Getting ready

Follow the *Hive batch mode - using a query file* recipe of *Chapter 6, Hadoop Ecosystem – Apache Hive* to create the Hive table that we'll be using in this recipe.

## How to do it...

This section demonstrates how to access a Hive table from Pig. Proceed with the following steps:

1. Start the Hive shell with the following command:

   ```
   $ hive
   ```

2. Issue the following command in the Hive shell to create the HBase table. The `HBaseStorageHandler` class takes care of the data communication with HBase. We have to specify the `hbase.column.mapping` property to instruct Hive on how to map the columns of the HBase table into the corresponding Hive table:

   ```
   CREATE TABLE IF NOT EXISTS books_hbase
     (key STRING,
     title STRING,
     author STRING,
     year INT,
     publisher STRING,
     image_s STRING,
     image_m STRING,
     image_l STRING)
   STORED BY 'org.apache.hadoop.hive.hbase.HBaseStorageHandler'
   WITH SERDEPROPERTIES ('hbase.columns.mapping' =
               ':key,f:title,f:author,f:year,f:publisher,
                img:image_s,img:image_m,img:image_l')
   TBLPROPERTIES ('hbase.table.name' = 'bx-books');
   ```

3. Issue the following Hive query to insert data into the newly created HBase table. RowKeys in HBase tables have to be unique. When there is more than one row with duplicate RowKeys, HBase stores only one of them and discards the others. Use the book ISBN, which is unique for each book, as the RowKey in the following example:

   ```
   hive> insert into table books_hbase
         select * from bookcrossing.books;
   ....
   Total MapReduce CPU Time Spent: 23 seconds 810 msec
   OK
   books.isbn    books.title    books.author    books.year    books.
   publisher    books.image_s    books.image_m    books.image_l
   Time taken: 37.817 seconds
   ```

4. Use the following command to inspect the data inserted to the Hive mapped HBase table:

   ```
   hive> select * from books_hbase limit 10;
   ```

5. We can also perform Hive functions, such as `count`, on the table we just created, as follows:

```
hive> select count(*) from books_hbase;
...
Total MapReduce CPU Time Spent: 22 seconds 510 msec
OK
_c0
271379
```

6. Start the HBase shell and issue the `list` command to see the list of tables in HBase, as follows:

```
$ hbase shell
hbase(main):001:0> list
TABLE
......
SYSTEM.STATS
bx-books
......
8 row(s) in 1.4260 seconds
```

7. Inspect the data of the `bx-books` HBase table using the following command:

```
hbase(main):003:0> scan 'bx-books', {'LIMIT' => 5}
```

The output of the preceding command is as follows:

```
hbase(main):003:0> scan 'bx-users', {'LIMIT' => 5}
ROW                      COLUMN+CELL
 0000913154               column=f:author, timestamp=1422502473588, value=C.
 0000913154               column=f:publisher, timestamp=1422502473588, value
 0000913154               column=f:title, timestamp=1422502473588, value=The
                          Technology
 0000913154               column=f:year, timestamp=1422502473588, value=1967
 0000913154               column=img:image_l, timestamp=1422502473588, value
                          .LZZZZZZZ.jpg
 0000913154               column=img:image_m, timestamp=1422502473588, value
                          .MZZZZZZZ.jpg
 0000913154               column=img:image_s, timestamp=1422502473588, value
                          .THUMBZZZ.jpg
 0001010565               column=f:author, timestamp=1422502471798, value=Ju
 0001010565               column=f:publisher, timestamp=1422502471798, value
 0001010565               column=f:title, timestamp=1422502471798, value=Mod
 0001010565               column=f:year, timestamp=1422502471798, value=1992
 0001010565               column=img:image_l, timestamp=1422502471798, value
                          .LZZZZZZZ.jpg
 0001010565               column=img:image_m, timestamp=1422502471798, value
                          .MZZZZZZZ.jpg
 0001010565               column=img:image_s, timestamp=1422502471798, value
                          .THUMBZZZ.jpg
```

▶ The *HCatalog – performing Java MapReduce computations on data mapped to Hive tables* and *HCatalog – writing data to Hive tables from Java MapReduce computations* recipes of *Chapter 6, Hadoop Ecosystem – Apache Hive*.

# Getting started with Apache Mahout

Mahout is an effort to implement well-known **machine learning** and **data mining** algorithms using the Hadoop MapReduce framework. Users can use Mahout algorithm implementations in their data processing applications without going through the complexity of implementing these algorithms using Hadoop MapReduce from scratch.

This recipe explains how to get started with Mahout.

In order to install Mahout, we recommend you use one of the freely available commercial Hadoop distributions as described in  *Chapter 1, Getting Started with Hadoop v2*. Another alternative is to use Apache Bigtop to install Mahout. Refer to the Bigtop-related recipe in *Chapter 1, Getting Started with Hadoop v2* for steps on installing Mahout using the Apache Bigtop distribution.

## How to do it...

This section demonstrates how to get started with Mahout by running a sample KMeans Clustering computation. You can run and verify the Mahout installation by carrying out the following steps:

1. Download the input data from `http://archive.ics.uci.edu/ml/databases/synthetic_control/synthetic_control.data` as follows:

```
$ wget http://archive.ics.uci.edu/ml/databases/synthetic_control/
synthetic_control.data
```

2. Create an HDFS directory named `testdata` and copy the downloaded file to that directory using the following command:

```
$ hdfs dfs -mkdir testdata
$ hdfs dfs -copyFromLocal synthetic_control.data  testdata
```

3. Run the K-mean sample by running the following command:

```
$ mahout org.apache.mahout.clustering.syntheticcontrol.kmeans.Job
```

4.  If all goes well, it will process and print out the clusters:

    ```
    12/06/19 21:18:15 INFO kmeans.Job: Running with default arguments
    12/06/19 21:18:15 INFO kmeans.Job: Preparing Input
    12/06/19 21:18:15 WARN mapred.JobClient: Use GenericOptionsParser
    for parsing the arguments. Applications should implement Tool for
    the same.

    . . . . .

    2/06/19 21:19:38 INFO clustering.ClusterDumper: Wrote 6 clusters
    12/06/19 21:19:38 INFO driver.MahoutDriver: Program took 83559 ms
    (Minutes: 1.39265)
    ```

## How it works...

Mahout is a collection of MapReduce jobs and you can run them using the `mahout` command. The preceding instructions installed and verified Mahout by running a **K-means** sample that comes with the Mahout distribution.

## There's more...

In addition to the next recipe in this chapter, the following recipes in *Chapter 10, Mass Text Data Processing* of this book also use Mahout:

▶  The *Creating TF and TF-IDF vectors for the text data* recipe of *Chapter 10, Mass Text Data Processing*

▶  The *Clustering text data using Apache Mahout* recipe of *Chapter 10, Mass Text Data Processing*

▶  The *Topic discovery using Latent Dirichlet Allocation (LDA)* recipe of *Chapter 10, Mass Text Data Processing*

▶  The *Document classification using Mahout Naive Bayes Classifier* recipe of *Chapter 10, Mass Text Data Processing*

# Running K-means with Mahout

K-means is a clustering algorithm. A clustering algorithm takes data points defined in an **N-dimensional space** and groups them into multiple **clusters** by considering the distance between those data points. A cluster is a set of data points such that the distance between the data points inside the cluster is much less than the distance from data points within cluster to data points outside the cluster. More details about the K-means clustering can be found from lecture 4 (`http://www.youtube.com/watch?v=1ZDybXl212Q`) of the *Cluster computing and MapReduce* lecture series by Google.

In this recipe, we will use a dataset that includes the **Human Development Report** (**HDR**) by country. The HDR describes different countries based on several human development measures. You can find the dataset at `http://hdr.undp.org/en/statistics/data/`. A sample of this dataset is available in the `chapter7/resources/hdi-data.csv` file in the sample source code repository. This recipe will use K-means to cluster countries based on HDR dimensions.

## Getting ready

This recipe needs a Mahout installation. Follow the previous recipe to install Mahout, if you haven't already done so.

## How to do it...

This section demonstrates how to use the Mahout K-means algorithm to process a dataset. Proceed with the following steps:

1.  Use the following Gradle command to compile the sample:

    ```
    $ gradle build
    ```

2.  Copy the file, `chapter7/resources/countries4Kmean.data`, to the `testdata` directory in HDFS. Create the `testdata` directory.

3.  Run the sample by running the following command:

    ```
    $ gradle executeKMeans
    ```

## How it works...

The preceding sample shows how you can configure and use K-means implementation from Java. You can find the source of this sample in the `chapter7/src/chapter7/KMeansSample.java` file. When we run the code, it initializes the K-means MapReduce job and executes it using the MapReduce framework.

# Importing data to HDFS from a relational database using Apache Sqoop

Apache Sqoop is a project that enables efficient bulk transfer of data between Apache Hadoop ecosystem and relational data stores. Sqoop can be used to automate the process of importing data from or exporting data to RDBMSs such as MySQL, PostgreSQL, Oracle, and so on. Sqoop also supports database appliances such as Netezza and Teradata, as well. It supports parallel import/export of data using multiple Map tasks and also supports throttling to reduce the load on the external RDBMSs.

In this recipe, we'll be using Sqoop2 to import data from a PostgreSQL database in to HDFS. We also include instructions for Sqoop 1.4.x as well, due to the wide availability and usage of that Sqoop version in the current Hadoop distributions.

We recommend that you use one of the freely available commercial Hadoop distributions as described in *Chapter 1, Getting Started with Hadoop v2*, to install Apache Sqoop2 or Sqoop 1.4.x. Another alternative is to use Apache Bigtop to install Apache Sqoop2.

## Getting ready

A working Hadoop2 cluster with a Sqoop2 or Sqoop 1.4.x installation is required for this recipe.

We will be using a PostgreSQL database. You can also use another RDBMS for this purpose, but certain steps of the following recipe will have to be changed accordingly.

## How to do it...

This section demonstrates how to import data from a PostgreSQL database in to HDFS using SQOOP. Proceed with the following steps:

1.  Download the appropriate PostgreSQL JDBC driver from `http://jdbc.postgresql.org/download.html` and copy it to the lib directory of the SQOOP web app using the following command and restart the SQOOP server:

    ```
    $ cp postgresql-XXXX.jdbcX.jar \
    /usr/lib/sqoop/webapps/sqoop/WEB-INF/lib/
    ```

     For Sqoop 1.4.x, copy the PostgreSQL JDBC driver jar to the lib folder of the Sqoop installation.

2.  Create an user and a database in the PostgreSQL, as follows. Use your OS username as the user in the PostgreSQL database as well. For this recipe, you can use an existing PostgreSQL user and an existing database as well:

    ```
    $ sudo su - postgres
    $ psql
    postgres=# CREATE USER aluck WITH PASSWORD 'xxx123';
    CREATE ROLE
    postgres=# CREATE DATABASE test;
    CREATE DATABASE
    postgres=# GRANT ALL PRIVILEGES ON DATABASE test TO aluck;
    GRANT
    postgres=# \q
    ```

3. Log in to the newly created database. Create a schema and a database table using the following statements in the PostgreSQL shell:

```
$ psql test
```

```
test=> CREATE SCHEMA bookcrossing;
CREATE SCHEMA
test=> CREATE TABLE bookcrossing.ratings
       (user_id INT,
        isbn TEXT,
        rating TEXT);
CREATE TABLE
```

4. Load the `book-ratings.txt` dataset in the `chapter7` folder of the Git repository into the table we just created, using the following command:

```
test=> \COPY bookcrossing.ratings FROM '…/chapter7/book-ratings.txt' DELIMITER ';'
test=# select * from bookcrossing.ratings limit 10;
```

```
 user_id |    isbn     | rating
---------+-------------+--------
  276725 | 034545104X  | 0
  276726 | 0155061224  | 5
  276727 | 0446520802  | 0
  276729 | 052165615X  | 3
  276729 | 0521795028  | 6
  276733 | 2080674722  | 0
  276736 | 3257224281  | 8
  276737 | 0600570967  | 6
  276744 | 038550120X  | 7
  276745 | 342310538   | 10
(10 rows)
```

Following steps (6 to 9) are for Sqoop2. Skip to step 10 for instructions on Sqoop 1.4.x.

5.  Create a SQOOP connection using the following command in the SQOOP command line client and answer the prompted questions:

```
$ sqoop
sqoop:000> create connection --cid 1
Creating connection for connector with id 1
Please fill following values to create new connection object
Name: t2


Connection configuration


JDBC Driver Class: org.postgresql.Driver
JDBC Connection String: jdbc:postgresql://localhost:5432/test
Username: testuser
Password: ****
JDBC Connection Properties:
There are currently 0 values in the map:
...

New connection was successfully created with validation status
FINE and persistent id 3
```

*[handwritten: Com. MYSQL, JDBC, DRIVER]*

*[handwritten: 33db]*

6.  Create a SQOOP job to import data into HDFS, as follows:

```
sqoop:000> create job --xid 1 --type import
Creating job for connection with id 1
Please fill following values to create new job object
Name: importest
Database configuration
Schema name: bookcrossing
Table name: ratings
Table SQL statement:
Table column names:
Partition column name: user_id
Boundary query:


Output configuration
Storage type:
    0 : HDFS
```

```
Choose: 0

Output format:
    0 : TEXT_FILE
    1 : SEQUENCE_FILE

Choose: 0

Output directory: /user/test/book_ratings_import
```

New job was successfully created with validation status FINE  and persistent id 8

7. Submit the Sqoop job with the following command:

```
sqoop:000> submission start --jid 8

Submission details

Job id: 8

Status: BOOTING

Creation date: 2014-10-15 00:01:20 EDT
```

8. Monitor the job status using this command:

```
sqoop:000> submission status --jid 8

Submission details

Job id: 8

Status: SUCCEEDED

Creation date: 2014-10-15 00:01:20 EDT
```

9. Check the HDFS directory for the data. You can map this data to Hive tables for further querying. Next two steps are only for Sqoop 1.4.x. Skip them if you are using Sqoop 2.

10. Issue the following Sqoop command to import the data from PostgreSQL directly in to a Hive table. Substitute the PostgreSQL database IP address (or hostname), database port and  database username accordingly. After the successful execution of the following command, a folder named 'ratings' containing the data imported from PostgreSQL will be created in your HDFS home directory:

```
$ sqoop import \
--connect jdbc:postgresql://<ip_address>:5432/test \
--table ratings \
--username aluck -P \
--direct -- --schema bookcrossing
```

11. Issue the following Sqoop command to import the data from PostgreSQL in to your HDFS home directory. Substitute the PostgreSQL database IP address (or hostname), database port and database username accordingly. After the successful execution of the following command, a Hive table named 'ratings' containing the data imported from PostgreSQL will be created in your current Hive database:

```
$ sqoop import \
--connect jdbc:postgresql://<ip_address>:5432/test \
--table ratings \
--username aluck -P \
--hive-import \
--direct -- --schema bookcrossing
```

# Exporting data from HDFS to a relational database using Apache Sqoop

In this recipe, we'll be using Sqoop2 or Sqoop 1.4.x to export data from HDFS to a PostgreSQL database.

## Getting ready

A working Hadoop2 cluster with a Sqoop2 or Sqoop 1.4.x installation is required for this recipe.

We will be using a PostgreSQL database. You can also use another RDBMS for this purpose as well, but the following recipe steps will have to be changed accordingly.

Follow the previous recipe, *Importing data to HDFS from a relational database using Apache Sqoop*.

## How to do it...

This section demonstrates how to export data from HDFS to a PostgreSQL database using SQOOP. Proceed with the following steps:

1. Follow the step 1 of the previous *Importing data to HDFS from a relational database using Apache Sqoop* recipe to create a user and a database in the PostgreSQL database.

2. Create a database table using the following statements in the PostgreSQL shell:

```
$ psql test
test=> CREATE TABLE bookcrossing.ratings_copy
  (user_id INT,
  isbn TEXT,
  rating TEXT);
```

 Following steps (3 to 5) are for Sqoop2. Skip to step 6 for instructions on Sqoop 1.4.x.

3. Create a SQOOP job to export data from HDFS, as follows:

```
sqoop:000> create job --xid 1 --type export
Creating job for connection with id 1
Please fill following values to create new job object
Name: exporttest

Database configuration
Schema name: bookcrossing
Table name: ratings_copy
Table SQL statement:
Table column names:
Input configuration
Input directory: /user/test/book_ratings_import
Throttling resources
Extractors:
Loaders:
New job was successfully created with validation status FINE  and
persistent id 13
```

4. Submit the Sqoop job with the following command:

```
sqoop:000> submission start --jid 13
Submission details
Job id: 13
Status: BOOTING
  …..
```

5. Monitor the job status using this command. Skip to step 7:

```
sqoop:000> submission status --jid 13
Submission details
Job id: 13
Status: SUCCEEDED
```

6. This step is only for Sqoop 1.4.x. Reexecute the step 11 of the previous *Importing data to HDFS from a relational database using Apache Sqoop* recipe to make sure you have the "ratings" folder with the imported data in your HDFS home directory. Issue the following Sqoop command to export the data from HDFS directly in to the PostgreSQL table. Substitute the PostgreSQL database IP address (or hostname), database port, database username, export data source directory accordingly. Execution of this step will result in a Hadoop MapReduce job:

```
$ sqoop export \
--connect jdbc:postgresql://<ip_address>:5432/test \
--table ratings_copy \
--username aluck -P \
--export-dir /user/aluck/ratings
--input-fields-terminated-by ','
--lines-terminated-by '\n'
-- --schema bookcrossing
```

7. Log in to the PostgreSQL shell and check the imported data:

```
test=# select * from bookcrossing.ratings_copy limit 10;
 user_id |    isbn     | rating
---------+-------------+--------
  276725 | 034545104X  | 0
  276726 | 0155061224  | 5
  276727 | 0446520802  | 0
  276729 | 052165615X  | 3
  276729 | 0521795028  | 6
```

# 8

# Searching and Indexing

In this chapter, we will cover the following recipes:

- ▸ Generating an inverted index using Hadoop MapReduce
- ▸ Intradomain web crawling using Apache Nutch
- ▸ Indexing and searching web documents using Apache Solr
- ▸ Configuring Apache HBase as the backend data store for Apache Nutch
- ▸ Whole web crawling with Apache Nutch using a Hadoop/HBase cluster
- ▸ Elasticsearch for indexing and searching
- ▸ Generating the in-links graph for crawled web pages

## Introduction

MapReduce frameworks are well suited for large-scale search and indexing applications. In fact, Google came up with the original MapReduce framework specifically to facilitate the various operations involved with web searching. The Apache Hadoop project was also started as a subproject for the **Apache Nutch** search engine, before spawning off as a separate top-level project.

Web searching consists of fetching, indexing, ranking, and retrieval. Given the very large size of data, all these operations need to be scalable. In addition, the retrieval should be low latency as well. Typically, fetching is performed through web crawling, where the crawlers fetch a set of pages in the fetch queue, extract links from the fetched pages, add the extracted links back to the fetch queue, and repeat this process many times. Indexing parses, organizes, and stores the fetched data in a manner that is fast and efficient for querying and retrieval. Search engines perform offline ranking of the documents based on algorithms such as PageRank and real-time ranking of the results based on the query parameters.

In this chapter, we introduce several tools that you can use with Apache Hadoop to perform large-scale searching and indexing.

**Sample code**

The example code files for this book are available in GitHub at `https://github.com/thilg/hcb-v2`. The `chapter8` folder code repository contains the sample code for this chapter.

Sample codes can be compiled and built by issuing the `gradle build` command in the `chapter8` folder of the code repository. The project files for Eclipse IDE can be generated by running the `gradle eclipse` command in the main folder of the code repository. The project files for IntelliJ IDEA IDE can be generated by running the `gradle idea` command in the main folder of the code repository.

# Generating an inverted index using Hadoop MapReduce

Simple text searching systems rely on inverted index to look up the set of documents that contain a given word or a term. In this recipe, we implement a simple inverted index building application that computes a list of terms in the documents, the set of documents that contains each term, and the term frequency in each of the documents. Retrieval of results from an inverted index can be as simple as returning the set of documents that contains the given terms or can involve much more complex operations such as returning the set of documents ordered based on a particular ranking.

## Getting ready

You must have Apache Hadoop v2 configured and installed to follow this recipe. Gradle is needed for the compiling and building of the source code.

## How to do it...

In the following steps, we use a MapReduce program to build an inverted index for a text dataset:

1. Create a directory in HDFS and upload a text dataset. This dataset should consist of one or more text files.

```
$ hdfs dfs -mkdir input_dir
$ hdfs dfs -put *.txt input_dir
```

You can download the text versions of the Project Gutenberg books by following the instructions given at http://www.gutenberg.org/ wiki/Gutenberg:Information_About_Robot_Access_to_ our_Pages. Make sure to provide the filetypes query parameter of the download request as txt. Unzip the downloaded files. You can use the unzipped text files as the text dataset for this recipe.

2. Compile the source by running the gradle build command from the chapter 8 folder of the source repository.

3. Run the inverted indexing MapReduce job using the following command. Provide the HDFS directory where you uploaded the input data in step 2 as the first argument and provide an HDFS path to store the output as the second argument:

```
$ hadoop jar hcb-c8-samples.jar \
        chapter8.invertindex.TextOutInvertedIndexMapReduce \
        input_dir output_dir
```

4. Check the output directory for the results by running the following command. The output of this program will consist of the term followed by a comma-separated list of filename and frequency:

```
$ hdfs dfs -cat output_dir/*

ARE three.txt:1,one.txt:1,four.txt:1,two.txt:1,

AS three.txt:2,one.txt:2,four.txt:2,two.txt:2,

AUGUSTA three.txt:1,

About three.txt:1,two.txt:1,

Abroad three.txt:2,
```

5. We used the text outputting inverted indexing MapReduce program in step 3 for the clarity of understanding the algorithm. The chapter8/invertindex/ InvertedIndexMapReduce.java MapReduce program in the source folder of chapter8 repository outputs the inverted index using the Hadoop SequenceFiles and MapWritable class. This index is friendlier for machine processing and more efficient for storage. You can run this version of the program by substituting the command in step 3 with the following command:

```
$ hadoop jar hcb-c8-samples.jar \
        chapter8.invertindex.InvertedIndexMapReduce \
        input_dir seq_output_dir
```

## How it works...

The Map Function receives a chunk of an input document as the input and outputs the term and
`<docid, 1>` pair for each word. In the Map function, we first replace all the non-alphanumeric
characters from the input text value before tokenizing it as follows:

```
public void map(Object key, Text value, ……… {
  String valString = value.toString().replaceAll("[^a-zA-Z0-9]+"," ");
  StringTokenizer itr = new StringTokenizer(valString);
   StringTokenizer(value.toString());

  FileSplit fileSplit = (FileSplit) context.getInputSplit();
  String fileName = fileSplit.getPath().getName();
  while (itr.hasMoreTokens()) {
    term.set(itr.nextToken());
    docFrequency.set(fileName, 1);
    context.write(term, docFrequency);
  }
}
```

We use the `getInputSplit()` method of `MapContext` to obtain a reference to
`InputSplit` assigned to the current Map task. The `InputSplits` class for this
computation are instances of `FileSplit` due to the usage of a `FileInputFormat` based
`InputFormat`. Then we use the `getPath()` method of `FileSplit` to obtain the path of
the file containing the current split and extract the filename from it. We use this extracted
filename as the document ID when constructing the inverted index.

The `Reduce` function receives IDs and frequencies of all the documents that contain the term
(Key) as the input. The `Reduce` function then outputs the term and a list of document IDs and
the number of occurrences of the term in each document as the output:

```
public void reduce(Text key, Iterable<TermFrequencyWritable>
values,Context context) …………{

  HashMap<Text, IntWritable> map = new HashMap<Text, IntWritable>();
  for (TermFrequencyWritable val : values) {
    Text docID = new Text(val.getDocumentID());
    int freq = val.getFreq().get();
    if (map.get(docID) != null) {
      map.put(docID, new IntWritable(map.get(docID).get() + freq));
    } else {
      map.put(docID, new IntWritable(freq));
    }
  }
  MapWritable outputMap = new MapWritable();
  outputMap.putAll(map),
  context.write(key, outputMap);
}
```

In the preceding model, we output a record for each word, generating a large amount of intermediate data between Map tasks and Reduce tasks. We use the following combiner to aggregate the terms emitted by the Map tasks, reducing the amount of Intermediate data that needs to be transferred between Map and Reduce tasks:

```
public void reduce(Text key, Iterable<TermFrequencyWritable> values ......
{
  int count = 0;
  String id = "";
  for (TermFrequencyWritable val : values) {
    count++;
    if (count == 1) {
      id = val.getDocumentID().toString();
    }
  }
  TermFrequencyWritable writable = new TermFrequencyWritable();
  writable.set(id, count);
  context.write(key, writable);
}
```

In the driver program, we set the Mapper, Reducer, and the Combiner classes. Also, we specify both Output Value and the MapOutput Value properties as we use different value types for the Map tasks and the reduce tasks.

```
...
job.setMapperClass(IndexingMapper.class);
job.setReducerClass(IndexingReducer.class);
job.setCombinerClass(IndexingCombiner.class);
...
job.setMapOutputValueClass(TermFrequencyWritable.class);
job.setOutputValueClass(MapWritable.class);
job.setOutputFormatClass(SequenceFileOutputFormat.class);
```

## There's more...

We can improve this indexing program by performing optimizations such as filtering stop words, substituting words with word stems, storing more information about the context of the word, and so on, making indexing a much more complex problem. Luckily, there exist several open source indexing frameworks that we can use for indexing purposes. The later recipes of this chapter will explore indexing using Apache Solr and Elasticsearch, which are based on the Apache Lucene indexing engine.

The upcoming section introduces the usage of `MapFileOutputFormat` to store `InvertedIndex` in an indexed random accessible manner.

## Outputting a random accessible indexed InvertedIndex

Apache Hadoop supports a file format called **MapFile** that can be used to store an index into the data stored in SequenceFiles. MapFile is very useful when we need to random access records stored in a large SequenceFile. You can use the **MapFileOutputFormat** format to output MapFiles, which would consist of a SequenceFile containing the actual data and another file containing the index into the SequenceFile.

The `chapter8/invertindex/MapFileOutInvertedIndexMR.java` MapReduce program in the source folder of `chapter8` utilizes MapFiles to store a secondary index into our inverted index. You can execute that program by using the following command. The third parameter (`sample_lookup_term`) should be a word that is present in your input dataset:

```
$ hadoop jar hcb-c8-samples.jar \
      chapter8.invertindex.MapFileOutInvertedIndexMR \
      input_dir indexed_output_dir sample_lookup_term
```

If you check `indexed_output_dir`, you will be able to see folders named as `part-r-xxxxx` with each containing a `data` and an `index` file. We can load these indexes to MapFileOutputFormat and perform random lookups for the data. An example of a simple lookup using this method is given in the `MapFileOutInvertedIndexMR.java` program as follows:

```
MapFile.Reader[] indexReaders = MapFileOutputFormat.getReaders
                               (new Path(args[1]), getConf());
MapWritable value = new MapWritable();
Text lookupKey = new Text(args[2]);
// Performing the lookup for the values if the lookupKey
Writable map = MapFileOutputFormat.getEntry(indexReaders,
        new HashPartitioner<Text, MapWritable>(), lookupKey, value);
```

In order to use this feature, you need to make sure to disable Hadoop from writing a `_SUCCESS` file in the `output` folder by setting the following property. The presence of the `_SUCCESS` file might cause an error when using MapFileOutputFormat to lookup the values in the index:

```
job.getConfiguration().setBoolean
        ("mapreduce.fileoutputcommitter.marksuccessfuljobs", false);
```

## See also

▶ The *Creating TF and TF-IDF vectors for the text data* recipe in *Chapter 10, Mass Text Data Processing*.

# Intradomain web crawling using Apache Nutch

**Web crawling** is the process of visiting and downloading all or a subset of web pages on the Internet. Although the concept of crawling and implementing a simple crawler sounds simple, building a full-fledged crawler takes a great deal of work. A full-fledged crawler needs to be distributed, has to obey the best practices such as not overloading servers and obey `robots.txt`, performing periodic crawls, prioritizing the pages to crawl, identifying many formats of documents, and so on. Apache Nutch is an open source search engine that provides a highly scalable crawler. Apache Nutch offers features such as politeness, robustness, and scalability.

In this recipe, we are going to use Apache Nutch in the standalone mode for small-scale intradomain web crawling. Almost all the Nutch commands are implemented as Hadoop MapReduce applications as you would notice when executing steps 10 to 18 of this recipe. Nutch standalone executed these applications using the Hadoop in local mode.

This recipe builds on the instructions given at `http://wiki.apache.org/nutch/NutchTutorial`.

## Getting ready

Set the `JAVA_HOME` environmental variable. Install Apache Ant and add it to the `PATH` environmental variable.

## How to do it...

The following steps show you how to use Apache Nutch in standalone mode for small scale web crawling:

1. Apache Nutch standalone mode uses the HyperSQL database as the default data storage. Download HyperSQL from `http://sourceforge.net/projects/hsqldb/`. Unzip the distribution and go to the data directory:

   ```
   $ cd hsqldb-2.3.2/hsqldb
   ```

2. Start a HyperSQL database using the following command. The following database uses `data/nutchdb.*` as the database files and uses `nutchdb` as the database alias name. We'll be using this database alias name in the `gora.sqlstore.jdbc.url` property in step 7:

   ```
   $ java -cp lib/hsqldb.jar \
   org.hsqldb.server.Server \
   ```

```
--database.0 file:data/nutchdb \

--dbname.0 nutchtest

......

[Server@79616c7]: Database [index=0, id=0, db=file:data/nutchdb,
alias=nutchdb] opened sucessfully in 523 ms.

......
```

3. Download Apache Nutch 2.2.1 from `http://nutch.apache.org/` and extract it.

4. Go to the extracted directory, which we will refer as `NUTCH_HOME`. Change the `gora-core` dependency version to 0.2.1 and uncomment the `gora-sql` dependency by modifying the `Gora artifacts` section of the `ivy/ivy.xml` file as follows:

```
<!--=================-->
<!-- Gora artifacts -->
<!--=================-->
<dependency org="org.apache.gora" name="gora-core"
  rev="0.2.1" conf="*->default"/>


<dependency org="org.apache.gora" name="gora-sql"
  rev="0.1.1-incubating" conf="*->default" />
```

> You can also use a MySQL database as the backend database for the Nutch standalone mode web crawling by updating the necessary database configurations in the `Default SqlStore properties` section of the `conf/gora.properties` file. You'll also have to uncomment the `mysql-connector-java` dependency in the `Gora artifacts` section of the `ivy/ivy.xml` file.

5. Build Apache Nutch using the following command:

```
$ ant runtime
```

6. Ensure you have the following in the `NUTCH_HOME/runtime/local/conf/gora.properties` file. Provide the database alias name used in step 2:

```
##############################
# Default SqlStore properties #
##############################
gora.sqlstore.jdbc.driver=org.hsqldb.jdbc.JDBCDriver
gora.sqlstore.jdbc.url=jdbc:hsqldb:hsql://
  localhost/nutchtest
gora.sqlstore.jdbc.user=sa
```

7. Go to the `runtime/local` directory and run the `bin/nutch` command to verify the Nutch installation. A successful installation would print out the list of Nutch commands as follows:

```
$ cd runtime/local
$ bin/nutch
Usage: nutch COMMAND
where COMMAND is one of:…..
```

8. Add the following to `NUTCH_HOME/runtime/local/conf/nutch-site.xml`. You can give any name to the value of `http.agent.name`:

```
<configuration>
<property>
  <name>storage.data.store.class</name>
  <value>org.apache.gora.sql.store.SqlStore</value>
</property>
<property>
  <name>http.agent.name</name>
  <value>NutchCrawler</value>
</property>
<property>
  <name>http.robots.agents</name>
  <value>NutchCrawler,*</value>
</property>
</configuration>
```

9. You can restrict the domain names you wish to crawl by editing the `regex-urlfiler.txt` file located at `NUTCH_HOME/runtime/local/conf/`. For example, in order to restrict the domain to `http://apache.org`, replace the following line at `NUTCH_HOME/runtime/local/conf/regex-urlfilter.txt`:

```
# accept anything else
+.
```

10. Using the following regular expression:

```
+^http://([a-z0-9]*\.)*apache.org/
```

11. Create a directory named `urls` and create a file named `seed.txt` inside that directory. Add your seed URLs to this file. Seed URLs are used to start the crawling and would be pages that are crawled first. We use `http://apache.org` as the seed URL in the following example:

```
$ mkdir urls
$ echo http://apache.org/ > urls/seed.txt
```

12. Inject the seed URLs into the Nutch database using the following command:

```
$ bin/nutch inject urls/
InjectorJob: starting
InjectorJob: urlDir: urls
......
Injector: finished
```

13. Use the following command to verify the injection of the seeds to the Nutch database. TOTAL urls printed by this command should match the number of URLs you had in your seed.txt file. You can use the following command in the later cycles as well to get an idea about the number of web page entries in your database:

```
$ bin/nutch readdb  -stats
WebTable statistics start
Statistics for WebTable:
min score:  1.0
. . . .
TOTAL urls:  1
```

14. Use the following command to generate a fetch list from the injected seed URLs. This will prepare the list of web pages to be fetched in the first cycle of the crawling. Generation will assign a batch-id to the current generated fetch list that can be used in the subsequent commands:

```
$ bin/nutch generate -topN 1
GeneratorJob: Selecting best-scoring urls due for fetch.
GeneratorJob: starting
GeneratorJob: filtering: true
GeneratorJob: done
GeneratorJob: generated batch id: 1350617353-1356796157
```

15. Use the following command to fetch the list of pages prepared in step 12. This step performs the actual fetching of the web pages. The -all parameter is used to inform Nutch to fetch all the generated batches:

```
$ bin/nutch fetch -all
FetcherJob: starting
FetcherJob: fetching all
```

```
FetcherJob: threads: 10

......


fetching http://apache.org/

......


-activeThreads=0
FetcherJob: done
```

16. Use the following command to parse and extract the useful data from fetched web pages, such as the text content of the pages, metadata of the pages, the set of pages linked from the fetched pages and so on. We call the set of pages linked from a fetched page the out-links of that particular fetched page. Out-links data would be used to discover new pages to fetch as well as to rank pages using link analysis algorithms such as PageRank:

```
$ bin/nutch parse -all
ParserJob: starting

......

ParserJob: success
```

17. Execute the following command to update the Nutch database with the data extracted in the preceding step. This step includes updating the contents of the fetched pages as well as adding new entries of the pages discovered through the links contained in the fetched pages.

```
$ bin/nutch updatedb
DbUpdaterJob: starting

......

DbUpdaterJob: done
```

18. Execute the following command to generate a new fetch list using the information from the previously fetched data. The topN parameter limits the number of URLs generated for the next fetch cycle:

```
$ bin/nutch generate -topN 100
GeneratorJob: Selecting best-scoring urls due for fetch.
GeneratorJob: starting

......

GeneratorJob: done
GeneratorJob: generated batch id: 1350618261-1660124671
```

19. Fetch the new list, parse it, and update the database.

```
$ bin/nutch fetch -all

......

$ bin/nutch parse -all

......

$ bin/nutch updatedb

......
```

20. Repeat steps 17 and 18 till you get the desired number of pages or the desired depth from your starting URLs.

## See also

▶ The *Whole web crawling with Apache Nutch using a Hadoop/HBase cluster* and *Indexing and searching web documents using Apache Solr* recipes.

▶ Refer to `http://www.hsqldb.org/doc/2.0/guide/index.html` for more information on using HyperSQL.

# Indexing and searching web documents using Apache Solr

**Apache Solr** is an open source search platform that is part of the **Apache Lucene** project. It supports powerful full-text search, faceted search, dynamic clustering, database integration, rich document (for example, Word and PDF) handling, and geospatial search. In this recipe, we are going to index the web pages crawled by Apache Nutch for use by Apache Solr and use Apache Solr to search through those web pages.

## Getting ready

1. Crawl a set of web pages using Apache Nutch by following the *Intradomain web crawling using Apache Nutch* recipe

2. Solr 4.8 and later versions require JDK 1.7

## How to do it...

The following steps show you how to index and search your crawled web pages dataset:

1.  Download and extract Apache Solr from `http://lucene.apache.org/solr/`. We use Apache Solr 4.10.3 for the examples in this chapter. From here on, we call the extracted directory as `$SOLR_HOME`.

2.  Replace the `schema.xml` file located under `$SOLR_HOME/examples/solr/collection1/conf/` using the `schema.solr4.xml` file located under `$NUTCH_HOME/runtime/local/conf/` as follows:

    ```
    $ cp $NUTCH_HOME/conf/schema-solr4.xml \
          $SOLR_HOME/example/solr/collection1/conf/schema.xml
    ```

3.  Add the following configuration to `$SOLR_HOME/examples/solr/collection1/conf/schema.xml` under the `<fields>` tag:

    ```
    <fields>
      <field name="_version_" type="long" indexed="true"
    stored="true"/>
    ......
    </fields>
    ```

4.  Start Solr by executing the following command from the `example` directory under `$SOLR_HOME/`:

    ```
    $ java -jar start.jar
    ```

5.  Go to the URL `http://localhost:8983/solr` to verify the Apache Solr installation.

6.  Index the data fetched using Apache Nutch into Apache Solr by issuing the following command from the `$NUTCH_HOME/runtime/local` directory. This command pushes the data crawled by Nutch into Solr through the Solr web service interface:

    ```
    $ bin/nutch solrindex http://127.0.0.1:8983/solr/ -reindex
    ```

7.  Go to Apache Solr search UI at `http://localhost:8983/solr/#/collection1/query`. Enter a search term in the **q** textbox and click on **Execute Query**, as shown in the following screenshot:

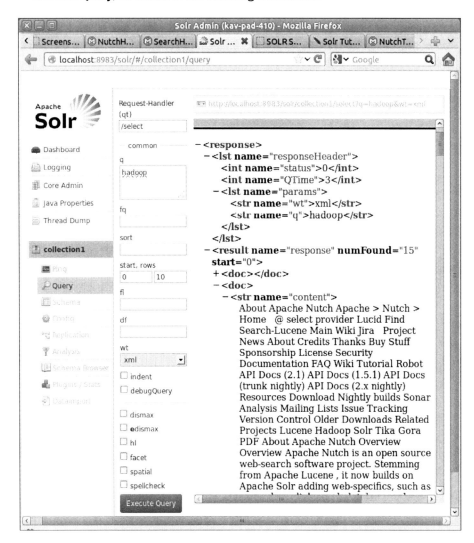

8.  You can also issue your search queries directly using the HTTP GET requests. Paste the `http://localhost:8983/solr/collection1/select?q=hadoop&start=5&rows=5&wt=xml` URL to your browser.

## How it works...

Apache Solr is built using the Apache Lucene text search library. Apache Solr adds many features on top of Apache Lucene and provides a text search web application that works out of the box. The preceding steps deploy Apache Solr and import the data crawled by Nutch into the deployed Solr instance.

The metadata of the documents we plan to index and search using Solr needs to be specified through the Solr `schema.xml` file. The Solr schema file should define the data fields in our documents and how these data fields should be processed by Solr. We use the schema file provided with Nutch (`$NUTCH_HOME/conf/schema-solr4.xml`), which defines the schema for the web pages crawled by Nutch, as the Solr schema file for this recipe. More information about the Solr schema file can be found at `http://wiki.apache.org/solr/SchemaXml`.

## See also

▶   The *Elasticsearch for indexing and searching* recipe.

▶   Follow the tutorial given at `http://lucene.apache.org/solr/tutorial.html` for more information on using Apache Solr.

▶   SolrCloud provides distributed indexing and searching capabilities for Apache Solr. More information on SolrCloud can be found at `https://cwiki.apache.org/confluence/display/solr/Getting+Started+with+SolrCloud`.

# Configuring Apache HBase as the backend data store for Apache Nutch

Apache Nutch integrates Apache Gora to add support for different backend data stores. In this recipe, we are going to configure Apache HBase as the backend data storage for Apache Nutch. Similarly, it is possible to plug in data stores such as RDBMS databases, Cassandra, and others through Gora.

This recipe builds upon the instructions given at `http://wiki.apache.org/nutch/Nutch2Tutorial`.

As of Apache Nutch 2.2.1 release, the Nutch project has not officially migrated to Hadoop 2.x and still depends on Hadoop 1.x for the whole web crawling. However, it is possible to execute the Nutch jobs using a Hadoop 2.x cluster utilizing the backward compatibility nature of Hadoop.

Nutch HBaseStore integration further depends on HBase 0.90.6, which doesn't support Hadoop 2. Hence, this recipe works only with a Hadoop 1.x cluster. We are looking forward to a new Nutch release with full Hadoop 2.x support.

## Getting ready

1. Install Apache Ant and add it to the `PATH` environmental variable.

## How to do it...

The following steps show you how to configure Apache HBase local mode as the backend data store for Apache Nutch to store the crawled data:

1. Install Apache HBase. Apache Nutch 2.2.1 and Apache Gora 0.3 recommend HBase 0.90.6 release.

2. Create two directories to store the HDFS data and Zookeeper data. Add the following to the `hbase-site.xml` file under `$HBASE_HOME/conf/` replacing the values with the paths to the two directories. Start HBase:

```
<configuration>
<property>
    <name>hbase.rootdir</name>
    <value>file:///u/software/hbase-0.90.6/hbase-data</value>
  </property>
<property>
    <name>hbase.zookeeper.property.dataDir</name>
    <value>file:///u/software/hbase-0.90.6/zookeeper-data</value>
  </property>
</configuration>
```

 Test your HBase installation using the HBase Shell before proceeding.

3. In case, you have not downloaded Apache Nutch for the earlier recipes in this chapter, download Nutch from `http://nutch.apache.org` and extract it.

4. Add the following to the `nutch-site.xml` file under `$NUTCH_HOME/conf/`:

```
<property>
 <name>storage.data.store.class</name>
 <value>org.apache.gora.hbase.store.HBaseStore</value>
 <description>Default class for storing data</description>
</property>
<property>
 <name>http.agent.name</name>
 <value>NutchCrawler</value>
</property>
```

```
<property>
  <name>http.robots.agents</name>
  <value>NutchCrawler,*</value>
</property>
```

5. Uncomment the following in the `Gora artifacts` section of the `ivy.xml` file under `$NUTCH_HOME/ivy/`. Revert the changes you made to the ivy/ivy.xml file in the earlier recipe and make sure that the `gora-core` dependency version is 0.3. Also, make sure to comment the `gora-sql` dependency:

```
<dependency org="org.apache.gora" name="gora-hbase"
  rev="0.3" conf="*->default" />
```

6. Add the following to the `gora.properties` file under `$NUTCH_HOME/conf/` to set the HBase storage as the default Gora data store:

```
gora.datastore.default=org.apache.gora.hbase.store.HBaseStore
```

7. Execute the following commands in the `$NUTCH_HOME` directory to build Apache Nutch with HBase as the backend data storage:

```
$ ant clean
$ ant runtime
```

8. Follow steps 9 to 19 of the *Intradomain web crawling using Apache Solr* recipe.

9. Start the Hbase shell and issue the following commands to view the fetched data:

```
$ hbase shell
HBase Shell; enter 'help<RETURN>' for list of supported commands.
Type "exit<RETURN>" to leave the HBase Shell
Version 0.90.6, r1295128, Wed Feb 29 14:29:21 UTC 2012
hbase(main):001:0> list
TABLE
webpage
1 row(s) in 0.4970 seconds

hbase(main):002:0> count 'webpage'
Current count: 1000, row: org.apache.bval:http/release-management.
html
Current count: 2000, row: org.apache.james:http/jspf/index.html
Current count: 3000, row: org.apache.sqoop:http/team-list.html
Current count: 4000, row: org.onesocialweb:http/
```

```
4065 row(s) in 1.2870 seconds


hbase(main):005:0> scan 'webpage',{STARTROW => 'org.apache.
nutch:http/', LIMIT=>10}
ROW                                 COLUMN+CELL
 org.apache.nutch:http/                 column=f:bas,
timestamp=1350800142780, value=http://nutch.apache.org/

 org.apache.nutch:http/                 column=f:cnt,
timestamp=1350800142780, value=<....

......

10 row(s) in 0.5160 seconds
```

10. Follow the steps in the *Indexing and searching web documents using Apache Solr* recipe and search the fetched data using Apache Solr.

## How it works...

The preceding steps configure and run Apache Nutch using Apache HBase as the storage backend. When configured, Nutch stores the fetched web page data and other metadata in HBase tables. In this recipe, we use a standalone HBase deployment. However, as shown in the *Whole web crawling with Apache Nutch using a Hadoop/HBase cluster* recipe, Nutch can be used with a distributed HBase deployment as well. Usage of HBase as the backend data store provides more scalability and performance for Nutch crawling.

## See also

▶ HBase recipes in *Chapter 7, Hadoop Ecosystem II – Pig, HBase, Mahout, and Sqoop*.

▶ Refer to http://techvineyard.blogspot.com/2010/12/build-nutch-20. html for instructions on configuring Cassandra or MySql as the storage backend for Nutch.

# Whole web crawling with Apache Nutch using a Hadoop/HBase cluster

Crawling a large amount of web documents can be done efficiently by utilizing the power of a MapReduce cluster.

As of Apache Nutch 2.2.1 release, the Nutch project has not officially migrated to Hadoop 2.x and still depends on Hadoop 1.x for the whole web crawling. However, it is possible to execute the Nutch jobs using a Hadoop 2.x cluster utilizing the backward compatibility nature of Hadoop.

Nutch HBaseStore integration further depends on HBase 0.90.6, which doesn't support Hadoop 2. Hence, this recipe works only with a Hadoop 1.x cluster. We are looking forward to a new Nutch release with full Hadoop 2.x support.

## Getting ready

We assume you already have your Hadoop 1.x and HBase cluster deployed.

## How to do it...

The following steps show you how to use Apache Nutch with a Hadoop MapReduce cluster and an HBase data store to perform large-scale web crawling:

1. Make sure the `hadoop` command is accessible from the command line. If not, add the `$HADOOP_HOME/bin` directory to the `PATH` environmental variable of your machine as follows:

   ```
   $ export PATH=$PATH:$HADOOP_HOME/bin/
   ```

2. Follow steps 3 to 7 of the *Configuring Apache HBase as the backend data store for Apache Nutch* recipe. You can skip this step if you have already followed that recipe.

3. Create a directory in HDFS to upload the seed urls.

   ```
   $ hadoop dfs -mkdir urls
   ```

4. Create a text file with the seed URLs for the crawl. Upload the seed URLs file to the directory created in the preceding step.

   ```
   $ hadoop dfs -put seed.txt urls
   ```

You can use the Open Directory project RDF dump (`http://rdf.dmoz.org/`) to create your seed URLs. Nutch provides a utility class to select a subset of URLs from the extracted DMOZ RDF data as `bin/nutch org.apache.nutch.tools.DmozParser content.rdf.u8 -subset 5000 > dmoz/urls`.

5. Issue the following command from $NUTCH_HOME/runtime/deploy to inject the seed URLs to the Nutch database and to generate the initial fetch list:

```
$ bin/nutch inject urls
$ bin/nutch generate
```

6. Issue the following commands from $NUTCH_HOME/runtime/deploy:

```
$ bin/nutch fetch -all
14/10/22 03:56:39 INFO fetcher.FetcherJob: FetcherJob: starting
14/10/22 03:56:39 INFO fetcher.FetcherJob: FetcherJob: fetching
all

......


$ bin/nutch parse -all
14/10/22 03:48:51 INFO parse.ParserJob: ParserJob: starting

......


14/10/22 03:50:44 INFO parse.ParserJob: ParserJob: success


$ bin/nutch updatedb
14/10/22 03:53:10 INFO crawl.DbUpdaterJob: DbUpdaterJob: starting
....
14/10/22 03:53:50 INFO crawl.DbUpdaterJob: DbUpdaterJob: done


$ bin/nutch generate -topN 10
14/10/22 03:51:09 INFO crawl.GeneratorJob: GeneratorJob: Selecting
best-scoring urls due for fetch.
14/10/22 03:51:09 INFO crawl.GeneratorJob: GeneratorJob: starting
....
14/10/22 03:51:46 INFO crawl.GeneratorJob: GeneratorJob: done
14/10/22 03:51:46 INFO crawl.GeneratorJob: GeneratorJob: generated
batch id: 1350892269-603479705
```

7. Repeat the commands in step 6 as many times as needed to crawl the desired number of pages or the desired depth.

8. Follow the *Indexing and searching web documents using Apache Solr* recipe to index the fetched data using Apache Solr.

## How it works...

All the Nutch operations we used in this recipe, including fetching and parsing, are implemented as MapReduce programs. These MapReduce programs utilize the Hadoop cluster to perform the Nutch operations in a distributed manner and use the HBase to store the data across the HDFS cluster. You can monitor these MapReduce computations through the monitoring UI of your Hadoop cluster.

Apache Nutch Ant build creates a Hadoop job file containing all the dependencies in the `deploy` folder under `$NUTCH_HOME/runtime/`. The `bin/nutch` script uses this job file to submit the MapReduce computations to the Hadoop cluster.

## See also

▶ The *Intradomain web crawling using Apache Nutch* recipe.

# Elasticsearch for indexing and searching

Elasticsearch (`http://www.elasticsearch.org/`) is an Apache 2.0 licensed open source search solution built on top of Apache Lucene. Elasticsearch is a distributed, multi-tenant, and document-oriented search engine. Elasticsearch supports distributed deployments, by breaking down an index into shards and by distributing the shards across the nodes in the cluster. While both Elasticsearch and Apache Solr use Apache Lucene as the core search engine, Elasticsearch aims to provide a more scalable and a distributed solution that is better suited for the cloud environments than Apache Solr.

## Getting ready

Install Apache Nutch and crawl some web pages as per the *Intradomain web crawling using Apache Nutch* or *Whole web crawling with Apache Nutch using a Hadoop/HBase cluster* recipe. Make sure the backend Hbase (or HyperSQL) data store for Nutch is still available.

## How to do it...

The following steps show you how to index and search the data crawled by Nutch using Elasticsearch:

1. Download and extract Elasticsearch from `http://www.elasticsearch.org/download/`.

2. Go to the extracted Elasticsearch directory and execute the following command to start the Elasticsearch server in the foreground:

   ```
   $ bin/elasticsearch
   ```

3. Run the following command in a new console to verify your installation:

   ```
   > curl localhost:9200
   {
      "status" : 200,
      "name" : "Talisman",
      "cluster_name" : "elasticsearch",
      "version" : {
      "number" : "1.4.2",
      ......
      "lucene_version" : "4.10.2"
      },
      "tagline" : "You Know, for Search"
   }
   ```

4. Go to the `$NUTCH_HOME/runtime/deploy` (or `$NUTCH_HOME/runtime/local` in case you are running Nutch in the local mode) directory. Execute the following command to index the data crawled by Nutch into the Elasticsearch server:

   ```
   $ bin/nutch elasticindex elasticsearch -all
   14/11/01 06:11:07 INFO elastic.ElasticIndexerJob: Starting
   ... . . .
   ```

5. Issue the following command to perform a search:

   ```
   $ curl -XGET 'http://localhost:9200/_search?q=hadoop'
   . . . .
   {"took":3,"timed_out":false,
     "_shards":{"total":5,"successful":5,"failed":0},
     "hits":{"total":36,"max_score":0.44754887,
       "hits":[{"_index":"index","_type":"doc","_id": 100 30551  100
         30551 "org.apache.hadoop:http/","_score":0.44754887,
   . . . .
   ```

## How it works...

Similar to Apache Solr, Elasticsearch too is built using the Apache Lucene text search library. In the preceding steps, we export the data crawled by Nutch into an instance of Elasticsearch for indexing and searching purposes.

You can also install Elasticsearch as a service as well. Refer to `http://www.elasticsearch.org/guide/reference/setup/installation.html` for more details on installing Elasticsearch as a service.

We use the ElasticIndex job of Nutch to import the data crawled by Nutch into the Elasticsearch server. Usage of the elasticindex command is as follows:

```
bin/nutch  elasticindex  <elastic cluster name> \
    (<batchId> | -all | -reindex) [-crawlId <id>]
```

The elastic cluster name is reverted to the default that is elasticsearch. You can change the cluster name by editing the `cluster.name` property in the `elasticsearch.yml` file under `config/`. Cluster name is used for autodiscovery purposes and should be unique for each Elasticsearch deployment in a single network.

## See also

▶ The *Indexing and searching web documents using Apache Solr* recipe.

# Generating the in-links graph for crawled web pages

The number of links to a particular web page from other pages, the number of in-links, is widely considered a good metric to measure the popularity or the importance of a web page. In fact, the number of in-links to a web page and the importance of the sources of those links have become integral components of most of the popular link analysis algorithms such as PageRank introduced by Google.

In this recipe, we are going to extract the in-links information from a set of web pages fetched by Apache Nutch and stored in Apache HBase backend data store. In our MapReduce program, we first retrieve the out-links information for the set of web pages stored in the Nutch HBase data store and then use that information to calculate the in-links graph for this set of web pages. The calculated in-link graph will contain the link information from the fetched subset of the web graph only.

## Getting ready

Follow the *Whole web crawling with Apache Nutch using a Hadoop/HBase cluster* recipe or the *Configuring Apache HBase as the backend data store for Apache Nutch* recipe and crawl a set of web pages using Apache Nutch to the backend HBase data store.

## How to do it...

The following steps show you how to extract out-links graph from the web pages stored in Nutch HBase data store and how to calculate the in-links graph using that extracted out-links graph:

1. Start the HBase shell:

   ```
   $ hbase shell
   ```

2. Create an HBase table with the name `linkdata` and a column family named `il`. Exit the HBase shell:

   ```
   hbase(main):002:0> create 'linkdata','il'

   0 row(s) in 1.8360 seconds

   hbase(main):002:0> quit
   ```

3. Unzip the source package for this chapter and compile it by executing `gradle build` from the `chapter8` source directory.

4. Run the Hadoop program by issuing the following command:

   ```
   $ hadoop jar hcb-c8-samples.jar \
   chapter8.InLinkGraphExtractor
   ```

5. Start the HBase shell and scan the `linkdata` table using the following command to check the output of the MapReduce program:

   ```
   $ hbase shell
   hbase(main):005:0> scan 'linkdata',{COLUMNS=>'il',LIMIT=>10}
   ROW                          COLUMN+CELL

   ....
   ```

# How it works...

Since we are going to use HBase to read input as well as to write the output, we use the HBase `TableMapper` and `TableReducer` helper classes to implement our MapReduce application. We configure the `TableMapper` and `TableReducer` classes using the utility methods given in the `TableMapReduceUtil` class. The `Scan` object is used to specify the criteria to be used by the mapper when reading the input data from the HBase data store:

```
Configuration conf = HBaseConfiguration.create();
Job job = new Job(conf, "InLinkGraphExtractor");
job.setJarByClass(InLinkGraphExtractor.class);
Scan scan = new Scan();
scan.addFamily("ol".getBytes());
TableMapReduceUtil.initTableMapperJob("webpage", scan, ......);
TableMapReduceUtil.initTableReducerJob("linkdata",......);
```

The map implementation receives the HBase rows as the input records. In our implementation, each of the rows corresponds to a fetched web page. The input key to the `Map` function consists of the web page URL and the value consists of the web pages linked from this particular web page. The `Map` function emits a record for each of the linked web pages, where the key of a `Map` output record is the URL of the linked page and the value of a `Map` output record is the input key to the `Map` function (the URL of the current processing web page):

```
public void map(ImmutableBytesWritable row, Result values,......){
  List<KeyValue> results = values.list();
  for (KeyValue keyValue : results) {
    ImmutableBytesWritable userKey = new
        ImmutableBytesWritable(keyValue.getQualifier());
    try {
      context.write(userKey, row);
    } catch (InterruptedException e) {
      throw new IOException(e);
    }
  }
}
```

The reduce implementation receives a web page URL as the key and a list of web pages that contain links to that web page (provided in the key) as the values. The reduce function stores this data into an HBase table:

```
public void reduce(ImmutableBytesWritable key,
    Iterable<ImmutableBytesWritable> values, ......{

Put put = new Put(key.get());
    for (ImmutableBytesWritable immutableBytesWritable :values)    {
        put.add(Bytes.toBytes("il"), Bytes.toBytes("link"),
            immutableBytesWritable.get());
    }
    context.write(key, put);
}
```

## See also

- ▶ The *Running MapReduce jobs on HBase* recipe in *Chapter 7, Hadoop Ecosystem II – Pig, HBase, Mahout, and Sqoop.*

# 9

# Classifications, Recommendations, and Finding Relationships

In this chapter, we will cover:

- ▶ Performing content-based recommendations
- ▶ Classification using the naïve Bayes classifier
- ▶ Assigning advertisements to keywords using the Adwords balance algorithm

# Introduction

This chapter discusses how we can use Hadoop for more complex use cases like classifying a dataset and making recommendations.

The following are a few instances of some such scenarios:

- ▶ Making product recommendations to users either based on similarities between products (for example, if a user liked a book about history, he/she might like another book on the same subject) or on user behavior patterns (for example, if two users are similar, they might like books the other has read)
- ▶ Clustering a dataset to identify similar entities; for example, identifying users with similar interests
- ▶ Classifying data into several groups based on historical data

In this recipe, we will apply these and other techniques using MapReduce. For recipes in this chapter, we will use the Amazon product co-purchasing network metadata dataset available at `http://snap.stanford.edu/data/amazon-meta.html`.

The contents of this chapter are based on *Chapter 8, Classifications, Recommendations, and Finding Relationships*, of the previous edition of this book, *Hadoop MapReduce Cookbook*. That chapter was contributed by the co-author Srinath Perera.

**Sample code**

The sample code and data files for this book are available in GitHub at `https://github.com/thilg/hcb-v2`. The `chapter9` folder of the code repository contains the sample source code files for this chapter. Sample codes can be compiled and built by issuing the `gradle build` command in the `chapter9` folder of the code repository. The project files for Eclipse IDE can be generated by running the `gradle eclipse` command in the main folder of the code repository. The project files for IntelliJ IDEA IDE can be generated by running the `gradle idea` command in the main folder of the code repository.

# Performing content-based recommendations

Recommendations are suggestions to someone about things that might be interesting to him. For example, you would recommend a good book to a friend who you know has similar interests as you. We often find use cases for recommendations in online retail. For example, when you browse a product, Amazon suggests other products also bought by users who bought that particular item.

An online retail site such as Amazon has a very large collection of items. Although books are found under several categories, often each category has too many to browse one after the other. Recommendations make the user's life easier by helping him find the best product for his tastes, and at the same time increase sales.

There are many ways to make recommendations:

- **Content-based recommendations**: One could use information about the product to identify similar products. For instance, you could use categories, content similarities, and so on, to identify products that are similar and recommend them to users who have already bought a particular product.

- **Collaborative filtering**: The other option is to use user behavior to identify similarities between products. For example, if the same user gave a high rating to two products, there is some similarity between those two products.

This recipe uses a dataset collected from Amazon about products to make content-based recommendations. In the dataset, each product has a list of similar items which is provided to the user, predetermined by Amazon. In this recipe, we will use that data to make recommendations.

## How to do it...

1. Download the dataset from the Amazon product co-purchasing network metadata available at `http://snap.stanford.edu/data/amazon-meta.html` and unzip it. We call this directory as `DATA_DIR`.

   Upload the data to HDFS by running the following commands. If the data directory is already there, clean it up.

   ```
   $ hdfs dfs -mkdir data
   $ hdfs dfs -mkdir data/input1
   $ hdfs dfs -put <DATA_DIR>/amazon-meta.txt data/input1
   ```

2. Compile the source by running the `gradle build` command from the `chapter9` directory of the source repository and obtain the `hcb-c9-samples.jar` file.

3. Run the most frequent user finder MapReduce job using the following command:

   ```
   $ hadoop jar hcb-c9-samples.jar \
     chapter9.MostFrequentUserFinder \
     data/input1 data/output1
   ```

4. Read the results by running the following command:

   ```
   $ hdfs dfs -cat data/output1/*
   ```

5. You will see that the MapReduce job has extracted the purchase data from each customer, and the results will look like the following:

   ```
   customerID=A1002VY75YRZYF,review=ASIN=0375812253#title=Really
   Useful Engines (Railway Series)#salesrank=623218#group=Book
   #rating=4#similar=0434804622|0434804614|0434804630|0679894780|
   0375827439|,review=ASIN=B000002BMD#title=EverythingMustGo#sale
   srank=77939#group=Music#rating=4#similar=B00000J5ZX|B000024J5H
   |B00005AWNW|B000025KKX|B000008I2Z
   ```

6. Run the recommendation MapReduce job using the following command:

   ```
   $ hadoop jar hcb-c9-samples.jar \
   chapter9.ContentBasedRecommendation \
   data/output1 data/output2
   ```

7.  Read the results by running the following command:

    ```
    $ hdfs dfs -cat data/output2/*
    ```

You will see that it will print the results as follows. Each line of the result contains the customer ID and a list of product recommendations for that customer.

```
A10003PM9DTGHQ  [0446611867, 0446613436, 0446608955, 0446606812,
0446691798, 0446611867, 0446613436, 0446608955, 0446606812, 0446691798]
```

## How it works...

The following listing shows an entry for one product from the dataset. Here, each data entry includes an ID, title, categorization, items similar to this item, and information about users who have reviewed the item. In this example, we assume that the customer who has reviewed the item has bought the item.

```
Id:    13
ASIN: 0313230269
   title: Clockwork Worlds : Mechanized Environments in SF
(Contributions to the Study of Science Fiction and Fantasy)
   group: Book
   salesrank: 2895088
   similar: 2  1559360968  1559361247
   categories: 3
      |Books[283155]|Subjects[1000]|Literature & Fiction[17]|History &
Criticism[10204]|Criticism & Theory[10207]|General[10213]
      |Books[283155]|Subjects[1000]|Science Fiction & Fantasy[25]|Fantasy
[16190]|History & Criticism[16203]
      |Books[283155]|Subjects[1000]|Science Fiction & Fantasy[25]|Science
Fiction[16272]|History & Criticism[16288]
   reviews: total: 2  downloaded: 2  avg rating: 5
      2002-8-5  cutomer: A14OJS0VWMOSWO  rating: 5  votes:   2  helpful:
1
      2003-3-21  cutomer: A2C27IQUH9N1Z  rating: 5  votes:   4
helpful:   4
```

We have written a Hadoop InputFormat to parse the Amazon product data; the data format works similar to the format we have written in the *Simple analytics using MapReduce* recipe of *Chapter 5, Analytics*. The source files, `src/chapter9/amazondata/AmazonDataReader.java` and `src/chapter9/amazondata/AmazonDataFormat.java`, contain the code for the Amazon data formatter.

The Amazon data formatter will parse the dataset and emit the data about each Amazon product as key-value pairs to the `map` function. Data about each Amazon product is represented as a string, and the `AmazonCustomer.java` class includes code to parse and write out the data about Amazon customers.

This recipe includes two MapReduce computations. The source for these tasks can be found from `src/chapter9/MostFrequentUserFinder.java` and `src/chapter9/ContentBasedRecommendation.java`. The Map task of the first MapReduce job receives data about each product in the log file as a different key-value pair.

When the Map task receives the product data, it emits the customer ID as the key and product information as the value for each customer who has bought the product.

```
public void map(Object key, Text value, Context context) throws
IOException, InterruptedException {
    List<AmazonCustomer> customerList =
    AmazonCustomer.parseAItemLine(value.toString());
    for(AmazonCustomer customer: customerList){
        context.write(new Text(customer.customerID),
        new Text(customer.toString()));
    }
}
```

Then, Hadoop collects all values for the key and invokes the Reducer once for each key. There will be a `reduce` function invocation for each customer, and each of those invocations will receive all products that have been bought by a customer. The Reducer emits the list of items bought by each customer, thus building a customer profile. Each of the items contains a list of similar items as well. In order to limit the size of the dataset, the Reducer will emit only the details of a customer who has bought more than five products.

```
public void reduce(Text key, Iterable<Text> values, Context
context) throws IOException, InterruptedException {
    AmazonCustomer  customer = new AmazonCustomer();
    customer.customerID = key.toString();

    for(Text value: values){
        Set<ItemData> itemsBought =new AmazonCustomer(
        value.toString()).itemsBought;
        for(ItemData itemData: itemsBrought){
            customer.itemsBought.add(itemData);
        }
    }
    if(customer.itemsBought.size() > 5){
        context.write(
        new IntWritable(customer.itemsBrought.size()),
        new Text(customer.toString()));
    }
}
```

The second MapReduce job uses the data generated from the first MapReduce task to make recommendations for each customer. The Map task receives data about each customer as the input, and the MapReduce job makes recommendations using the following three steps:

1. Each product (item) data from Amazon includes items similar to that item. Given a customer, the `map` function creates a list of all similar items for each item that customer has bought.

2. Then, the `map` function removes any item that the customer has already bought from the similar items list.

3. Finally, the `map` function selects ten items as recommendations.

```
public void map(Object key, Text value, Context context)
throws IOException, InterruptedException {
    AmazonCustomer amazonCustomer =
    new AmazonCustomer(value.toString()
    .replaceAll("[0-9]+\\s+", ""));

    List<String> recommendations = new ArrayList<String>();
    for (ItemData itemData : amazonCustomer.itemsBrought) {
        recommendations.addAll(itemData.similarItems);
    }

    for (ItemData itemData : amazonCustomer.itemsBrought) {
        recommendations.remove(itemData.itemID);
    }

    ArrayList<String> finalRecommendations =
    new ArrayList<String>();
    for (int i = 0;
    i < Math.min(10, recommendations.size());i++) {
        finalRecommendations.add(recommendations.get(i));
    }
    context.write(new Text(amazonCustomer.customerID),
    new Text(finalRecommendations.toString()));
}
```

## There's more...

You can learn more about content-based recommendations from *Chapter 9, Recommendation Systems*, of the book, *Mining of Massive Datasets, Cambridge University Press*, by Anand Rajaraman and Jeffrey David Ullman.

Apache Mahout, introduced in *Chapter 7, Hadoop Ecosystem II –Pig, HBase, Mahout, and Sqoop,* and used in *Chapter 10, Mass Text Data Processing,* contains several recommendation implementations. The following articles will give you information on using user-based and item-based recommenders in Mahout:

- https://mahout.apache.org/users/recommender/userbased-5-minutes.html

- https://mahout.apache.org/users/recommender/intro-itembased-hadoop.html

# Classification using the naïve Bayes classifier

A **classifier** assigns inputs into one of the *N* classes based on some properties (also known as features) of inputs. Classifiers have widespread applications, such as e-mail spam filtering, finding the most promising products, selecting customers for closer interactions, and taking decisions in machine learning situations. Let's explore how to implement a classifier using a large dataset. For instance, a spam filter will assign each e-mail to one of the two clusters: spam mail or not spam mail.

There are many classification algorithms. One of the simplest, but effective, algorithm is the naïve Bayesian classifier that uses the Bayes theorem involving conditional probability.

In this recipe, we will also focus on the Amazon metadata dataset as before. We will look at several features of a product, such as the number of reviews received, positive ratings, and known similar items to identify a product with potential to be within the first 10,000 sales rank. We will use the naïve Bayesian classifier for this classification.

 You can learn more about the naïve Bayer classifier at http://en.wikipedia.org/wiki/Naive_Bayes_classifier.

## How to do it...

1. Download the dataset from the Amazon product co-purchasing network metadata available at http://snap.stanford.edu/data/amazon-meta.html and unzip it. We will call this directory DATA_DIR.

2. Upload the data to HDFS by running the following commands. If the data directory is already there, clean it up.

   ```
   $ hdfs dfs -mkdir data
   $ hdfs dfs -mkdir data/input1
   $ hdfs dfs -put <DATA_DIR>/amazon-meta.txt data/input1
   ```

3. Compile the source by running the `gradle build` command from the `chapter9` directory of the source repository and obtain the `hcb-c9-samples.jar` file.

4. Run the MapReduce job using the following command:

```
$ hadoop jar hcb-c9-samples.jar \ chapter9.
NaiveBayesProductClassifier \
data/input1 data/output5
```

5. Read the results by running the following command:

```
$ hdfs dfs -cat data/output5/*
```

6. You will see that it will print the following results. You can use these values with the Bayes classifier to classify the inputs:

```
postiveReviews>30          0.635593220338983

reviewCount>60   0.62890625

similarItemCount>150       0.5720620842572062
```

## How it works...

The classifier uses the following features as indicators that the product can fall within the first 10,000 products:

▶ Number of reviews for a given product

▶ Number of positive reviews for a given product

▶ Number of similar items for a given product

We first run the MapReduce task to calculate the following probabilities, and then we will use those with the preceding formula to classify a given product:

▶ **P1**: Probability that a given item is within the first 10,000 products if it has more than 60 reviews

▶ **P2**: Probability that a given item is within the first 10,000 products if it has more than 30 positive reviews

▶ **P3**: Probability that a given item is within the first 10,000 products if it has more than 150 similar items

You can find the source for the classifier in the file, `src/chapter9/NaiveBayesProductClassifier.java`. The Mapper function looks like this:

```
public void map(Object key, Text value, Context context) throws
IOException, InterruptedException {
  List<AmazonCustomer> customerList =
    AmazonCustomer.parseAItemLine(value.toString());
```

```
    int salesRank = -1;
    int reviewCount = 0;
    int postiveReviews = 0;
    int similarItemCount = 0;

    for (AmazonCustomer customer : customerList) {
      ItemData itemData =
        customer.itemsBrought.iterator().next();
      reviewCount++;
      if (itemData.rating > 3) {
        postiveReviews++;
      }
      similarItemCount = similarItemCount +
        itemData.similarItems.size();
      if (salesRank == -1) {
        salesRank = itemData.salesrank;
      }
    }

    boolean isInFirst10k = (salesRank <= 10000);
    context.write(new Text("total"),
    new BooleanWritable(isInFirst10k));
    if (reviewCount > 60) {
      context.write(new Text("reviewCount>60"),
      new BooleanWritable(isInFirst10k));
    }
    if (postiveReviews > 30) {
      context.write(new Text("postiveReviews>30"),
      new BooleanWritable(isInFirst10k));
    }
    if (similarItemCount > 150) {
      context.write(new Text("similarItemCount>150"),
      new BooleanWritable(isInFirst10k));
    }
  }
```

The Mapper function walks though each product and evaluates its features. If the feature evaluates as true, it emits the feature name as the key and whether the product is within the first 10,000 products as the value.

MapReduce invokes the Reducer once for each feature. Then, each Reduce job receives all values for which the feature is true, and it calculates the probability that the product is within the first 10,000 products in the sales rank, given the feature is true.

```
public void reduce(Text key, Iterable<BooleanWritable> values,
Context context) throws IOException,
  InterruptedException {
    int total = 0;
    int matches = 0;
    for (BooleanWritable value : values) {
      total++;
      if (value.get()) {
        matches++;
      }
    }
  context.write(new Text(key),
    new DoubleWritable((double) matches / total));
  }
```

Given a product, we will examine and decide the following:

▶ Does it have more than 60 reviews?

▶ Does it have more than 30 positive reviews?

▶ Does it have more than 150 similar items?

We can decide the probabilities of events A, B, and C and we can calculate the probability of the given item being within the top 10,000 products using the Bayes theorem. The following code implements this logic:

```
public static boolean classifyItem(int similarItemCount, int
reviewCount, int postiveReviews){
  double reviewCountGT60 = 0.8;
  double postiveReviewsGT30 = 0.9;
  double similarItemCountGT150 = 0.7;
  double a , b, c;

  if (reviewCount > 60) {
    a = reviewCountGT60;
  }else{
    a= 1 - reviewCountGT60;
  }
  if (postiveReviews > 30) {
    b = postiveReviewsGT30;
  }else{
```

```
      b = 1- postiveReviewsGT30;
    }
    if (similarItemCount > 150) {
      c = similarItemCountGT150;
    }else{
      c = 1- similarItemCountGT150;
    }
    double p = a*b*c/ (a*b*c + (1-a)*(1-b)*(1-c));
    return p > 0.5;
  }
```

When you run the classifier testing logic, it will load the data generated by the MapReduce job and classify 1,000 randomly selected products.

# Assigning advertisements to keywords using the Adwords balance algorithm

Advertisements have become a major medium of revenue for the Web. It is a billion dollar business and is the source for revenue of most leading companies in Silicon Valley. Further, it has made it possible for companies such as Google, Facebook, Yahoo!, and Twitter to run their main services for free while collecting their revenue through advertisements.

Adwords lets people bid for keywords. For example, advertiser *A* can bid for the keyword, Hadoop Support, for $2 and provide a maximum budget of $100. Advertiser *B* can bid for the keyword, Hadoop Support, for $1.50 and provide a maximum budget of $200. When a user searches for a document with the given keywords, the system will choose one or more advertisements among the bids for these keywords. Advertisers will pay only if a user clicks on the advertisement.

The goal is to select advertisements such that they will maximize revenue. There are several factors in play when designing such a solution:

- We want to show advertisements that are more likely to be clicked often, as often times only clicks, not showing the advertisement, will get us money. We measure this as the fraction of times an advertisement was clicked as opposed to the number of times it was shown. We call this click-through rate for a keyword.

- We want to show advertisements belonging to advertisers with higher budgets as opposed to those with lower budgets.

In this recipe, we will implement a simplified version of the Adwords balance algorithm that can be used in such situations. For simplicity, we will assume that advertisers only bid on single words. Also, since we cannot find a real bid dataset, we will generate a sample bid dataset. Have a look at the following figure:

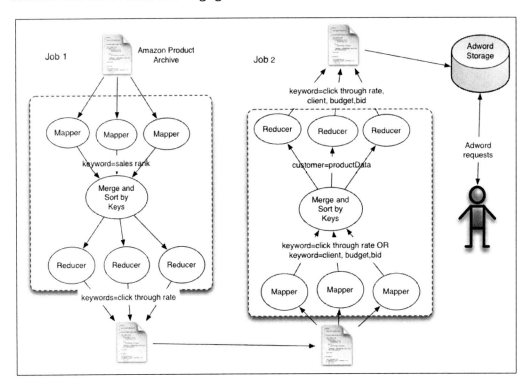

Assume that you are to support keyword-based advertisements using the Amazon dataset. The recipe will work as follows:

1. The first MapReduce job will approximate the click-through rate of the keyword using the Amazon sales index. Here, we assume that keywords that are found in the title of products with higher sales rank will have better click-through rates.

2. Then, we will run a Java program to generate a bid dataset.

3. Now, the second MapReduce task will group bids for the same product together and create an output that can be used by an advertisement assignment program.

4. Finally, we will use an advertisement assignment program to assign keywords to advertisers. We will use the following formula to implement the Adword balance algorithm. The formula assigns priority based on the fraction of unspent budget of each advertiser, bid value, and click-through rate:

```
Measure = bid value * click-through rate * (1-e^(-1*current
budget/ initial budget))
```

## How to do it...

1. Download the dataset from the Amazon product co-purchasing network metadata available from `http://snap.stanford.edu/data/amazon-meta.html` and unzip it. We will call this directory `DATA_DIR`.

2. Upload the data to HDFS by running the following commands. If data directory is already there, clean it up.

   ```
   $ hdfs dfs -mkdir data
   $ hdfs dfs -mkdir data/input1
   $ hdfs dfs -put <DATA_DIR>/amazon-meta.txt data/input1
   ```

3. Compile the source by running the `gradle build` command from the `chapter9` directory of the source repository and obtain the `hcb-c9-samples.jar` file.

4. Run the MapReduce job using the following command:

   ```
   $ hadoop jar hcb-c9-samples.jar \
   chapter9.adwords.ClickRateApproximator \
   data/input1 data/output6
   ```

5. Download the results to your computer by running the following command:

   ```
   $ hdfs dfs -get data/output6/part-r-* clickrate.data
   ```

6. You will see that it will print the following results. You may use these values with the Bayes classifier to classify the inputs:

   ```
   keyword:(Annals 74
   keyword:(Audio   153
   keyword:(BET     95
   keyword:(Beat    98
   keyword:(Beginners)     429
   keyword:(Beginning      110
   ```

7. Generate a bid dataset by running the following command. You can find the results in a `biddata.data` file.

   ```
   $ java -cp hcb-c9-samples.jar \
     chapter9.adwords.AdwordsBidGenerator \
     clickrate.data
   ```

8. Create a directory called `data/input2` and upload the bid dataset and results from the earlier MapReduce task to the `data/input2` directory of HDFS with the following command:

   ```
   $ hdfs dfs -put clickrate.data data/input2
   $ hdfs dfs -put biddata.data data/input2
   ```

9. Run the second MapReduce job as follows:

```
$ hadoop jar hcb-c9-samples.jar \
    chapter9.adwords.AdwordsBalanceAlgorithmDataGenerator \
    data/input2 data/output7
```

10. Download the results to your computer by running the following command:

```
$ hdfs dfs -get data/output7/part-r-* adwords.data
```

11. Inspect the results:

```
(Animated        client23,773.0,5.0,97.0|
(Animated)       client33,310.0,8.0,90.0|
(Annals          client76,443.0,13.0,74.0|
client51,1951.0,4.0,74.0|
(Beginners)      client86, 210.0,6.0,429.0|
    client6,236.0,5.0,429.0|
(Beginning       client31,23.0,10.0,110.0|
```

12. Perform matches for random sets of keywords by running the following command:

```
$ java jar hcb-c9-samples.jar \
    chapter9.adwords.AdwordsAssigner adwords.data
```

## How it works...

As we discussed, this recipe consists of two MapReduce jobs. You can find the source for the first MapReduce job from the file, `src/chapter9/adwords/ClickRateApproximator.java`.

The Mapper function parses the Amazon dataset using the Amazon data format, and for each word in each product title, it emits the word and the sales ranks of that product. The function looks something like this:

```
public void map(Object key, Text value, Context context) {
......
    String[] tokens = itemData.title.split("\\s");
    for(String token: tokens){
        if(token.length() > 3){
            context.write(new Text(token),
                    new IntWritable(itemData.salesrank));
        }
    }
}
```

Then, the MapReduce framework sorts the emitted key-value pairs by key and invokes the Reducer once for each key. As shown here, the reducer calculates an approximation for click rate using sales ranks emitted against the key:

```
public void reduce(Text key, Iterable<IntWritable> values, Context
context) throws IOException, InterruptedException {
    double clickrate = 0;
    for(IntWritable val: values){
        if(val.get() > 1){
            clickrate = clickrate + 1000/Math.log(val.get());
        }else{
            clickrate = clickrate + 1000;
        }
    }
    context.write(new Text("keyword:" +key.toString()),
    new IntWritable((int)clickrate));
}
```

There is no publicly available bid dataset. Therefore, we will generate a random bid dataset for our recipe using the `AdwordsBidGenerator` program. It will read the keywords generated by the preceding recipe and generate a random bid dataset.

Then, we will use the second MapReduce job to merge the bid dataset with the click-through rate and generate a dataset that has bid information sorted against the keyword. You can find the source for the second MapReduce task from the file, `src/chapter9/adwords/AdwordsBalanceAlgorithmDataGenerator.java`. The Mapper function looks like this:

```
public void map(Object key, Text value, Context context) throws
IOException, InterruptedException {
  String[] keyVal = value.toString().split("\\s");
  if (keyVal[0].startsWith("keyword:")) {
    context.write(
    new Text(keyVal[0].replace("keyword:", "")),
    new Text(keyVal[1]));
  } else if (keyVal[0].startsWith("client")) {
    List<String[]> bids = new ArrayList<String[]>();
    double budget = 0;
    String clientid = keyVal[0];
    String[] tokens = keyVal[1].split(",");
    for (String token : tokens) {
      String[] kp = token.split("=");
      if (kp[0].equals("budget")) {
        budget = Double.parseDouble(kp[1]);
      } else if (kp[0].equals("bid")) {
```

```
        String[] bidData = kp[1].split("\\|");
        bids.add(bidData);
      }
    }

  for (String[] bid : bids) {
    String keyword = bid[0];
    String bidValue = bid[1];
    Double.parseDouble(bidValue);
    context.write(new Text(keyword),
    new Text(new StringBuffer()
    .append(clientid).append(",")
    .append(budget).append(",")
    .append(bidValue).toString()));
  }
 }
}
```

The Mapper function reads both, the bid dataset and click-through rate dataset and
emits both types of data against the keyword. Then, each Reducer receives all bids and
the associated click-through data for each keyword. Next, the reducer merges the data
and emits a list of bids against each keyword.

```
public void reduce(Text key, Iterable<Text> values,
Context context) throws IOException, InterruptedException {
  String clientid = null;
  String budget = null;
  String bid = null;
  String clickRate = null;

  List<String> bids = new ArrayList<String>();
  for (Text val : values) {
    if (val.toString().indexOf(",") > 0) {
      bids.add(val.toString());
    } else {
      clickRate = val.toString();
    }
  }
  StringBuffer buf = new StringBuffer();
  for (String bidData : bids) {
    String[] vals = bidData.split(",");
```

```
    clientid = vals[0];
    budget = vals[1];
    bid = vals[2];
    buf.append(clientid).append(",")
    .append(budget).append(",")
    .append(Double.valueOf(bid)).append(",")
    .append(Math.max(1, Double.valueOf(clickRate)));
    buf.append("|");
  }
  if (bids.size() > 0) {
    context.write(key, new Text(buf.toString()));
  }
}
```

Finally, the Adwords assigner loads the bid data and stores it against keywords to memory. Given a keyword, the Adwords assigner finds the bid that has the maximum value for the following equation and selects a bid among all the bids for advertisement:

```
Measure = bid value * click-through rate * (1-e^(-1*current
budget/ initial budget))
```

## There's more...

The preceding recipe assumes that the Adwords assigner can load all the data into memory to make advertisement assignment decisions. In reality, these computations are handled by large clusters making real-time decisions combining streaming technologies such as Apache Storm and high-throughput databases such as HBase, due to the millisecond level response times and the large datasets required by advertisement bidding systems.

This recipe assumes that users only bid for single words. However, to support multiple keyword bids, we would need to combine the click-through rates, and the rest of the algorithm can proceed as earlier.

More information about online advertisement can be found in the book, *Mining of Massive Datasets, Cambridge University Press*, by Anand Rajaraman and Jeffrey David Ullman.

# 10

# Mass Text Data Processing

In this chapter, we will cover the following topics:

- Data preprocessing (extract, clean, and format conversion) using Hadoop streaming and Python
- De-duplicating data using Hadoop streaming
- Loading large datasets to an Apache HBase data store – importtsv and bulkload
- Creating TF and TF-IDF vectors for the text data
- Clustering text data using Apache Mahout
- Topic discovery using Latent Dirichlet Allocation (LDA)
- Document classification using Mahout Naive Bayes Classifier

## Introduction

Hadoop MapReduce together with the supportive set of projects makes it a good framework of choice to process large text datasets and to perform **extract-transform-load** (**ETL**) type operations.

In this chapter, we'll be exploring how to use Hadoop streaming to perform data preprocessing operations such as data extraction, format conversion, and de-duplication. We'll also use HBase as the data store to store the data and will explore mechanisms to perform large bulk data loads to HBase with minimal overhead. Finally, we'll look into performing text analytics using the Apache Mahout algorithms.

We will be using the following sample dataset for the recipes in this chapter:

- 20 Newsgroups dataset available at `http://qwone.com/~jason/20Newsgroups`. This dataset contains approximately 20,000 newsgroup documents originally collected by Ken Lang.

[

**Sample code**

The example code files for this book are available in GitHub at `https://github.com/thilg/hcb-v2`. The `chapter10` folder of the code repository contains the sample code for this chapter.
]

# Data preprocessing using Hadoop streaming and Python

Data preprocessing is an important and often required component in data analytics. Data preprocessing becomes even more important when consuming unstructured text data generated from multiple different sources. Data preprocessing steps include operations such as cleaning the data, extracting important features from data, removing duplicate items from the datasets, converting data formats, and many more.

Hadoop MapReduce provides an ideal environment to perform these tasks in parallel when processing massive datasets. Apart from using Java MapReduce programs or Pig scripts or Hive scripts to preprocess the data, Hadoop also contains several other tools and features that are useful in performing these data preprocessing operations. One such feature is the `InputFormats`, which provides us with the ability to support custom data formats by implementing custom `InputFormats`. Another feature is the Hadoop streaming support, which allows us to use our favorite scripting languages to perform the actual data cleansing and extraction, while Hadoop will parallelize the computation to hundreds of compute and storage resources.

In this recipe, we are going to use Hadoop streaming with a Python script-based Mapper to perform data extraction and format conversion.

## Getting ready

- Check whether Python is already installed on the Hadoop worker nodes. If not, install Python on all the Hadoop worker nodes.

## How to do it...

The following steps show how to clean and extract data from the 20news dataset and store the data as a tab-separated file:

1. Download and extract the 20news dataset from http://qwone.com/~jason/20N ewsgroups/20news-19997.tar.gz:

   ```
   $ wget http://qwone.com/~jason/20Newsgroups/20news-19997.tar.gz
   $ tar -xzf 20news-19997.tar.gz
   ```

2. Upload the extracted data to the HDFS. In order to save the compute time and resources, you can use only a subset of the dataset:

   ```
   $ hdfs dfs -mkdir 20news-all
   $ hdfs dfs -put  <extracted_folder> 20news-all
   ```

3. Extract the resource package for this chapter and locate the `MailPreProcessor.py` Python script.

4. Locate the `hadoop-streaming.jar` JAR file of the Hadoop installation in your machine. Run the following Hadoop streaming command using that JAR. `/usr/lib/hadoop-mapreduce/` is the `hadoop-streaming` JAR file's location for the Bigtop-based Hadoop installations:

   ```
   $ hadoop jar \
   /usr/lib/hadoop-mapreduce/hadoop-streaming.jar \
   -input 20news-all/*/* \
   -output 20news-cleaned \
   -mapper MailPreProcessor.py \
   -file MailPreProcessor.py
   ```

5. Inspect the results using the following command:

   ```
   > hdfs dfs -cat 20news-cleaned/part-*  | more
   ```

## How it works...

Hadoop uses the default `TextInputFormat` as the input specification for the previous computation. Usage of the `TextInputFormat` generates a Map task for each file in the input dataset and generates a Map input record for each line. Hadoop streaming provides the input to the Map application through the standard input:

```
line =  sys.stdin.readline();
while line:
....
```

```
    if (doneHeaders):
      list.append( line )
    elif line.find( "Message-ID:" ) != -1:
      messageID = line[ len("Message-ID:"):]
    ....
    elif line == "":
      doneHeaders = True

     line = sys.stdin.readline();
```

The preceding Python code reads the input lines from the standard input until it reaches the end of the file. We parse the headers of the newsgroup file till we encounter the empty line that demarcates the headers from the message contents. The message content will be read in to a list line by line:

```
value = ' '.join( list )
value = fromAddress + "\t" ......"\t" + value
print '%s\t%s' % (messageID, value)
```

The preceding code segment merges the message content to a single string and constructs the output value of the streaming application as a tab-delimited set of selected headers, followed by the message content. The output key value is the `Message-ID` header extracted from the input file. The output is written to the standard output by using a tab to delimit the key and the value.

## There's more...

We can generate the output of the preceding computation in the Hadoop `SequenceFile` format by specifying `SequenceFileOutputFormat` as the `OutputFormat` of the streaming computations:

```
$ hadoop jar \
/usr/lib/Hadoop-mapreduce/hadoop-streaming.jar \
-input 20news-all/*/* \
-output 20news-cleaned \
-mapper MailPreProcessor.py \
-file MailPreProcessor.py \
-outputformat \
        org.apache.hadoop.mapred.SequenceFileOutputFormat \
-file MailPreProcessor.py
```

It is a good practice to store the data as `SequenceFiles` (or other Hadoop binary file formats such as Avro) after the first pass of the input data because `SequenceFiles` takes up less space and supports compression. You can use `hdfs dfs -text <path_to_ sequencefile>` to output the contents of a `SequenceFile` to text:

```
$ hdfs dfs -text 20news-seq/part-* | more
```

However, for the preceding command to work, any Writable classes that are used in the `SequenceFile` should be available in the Hadoop classpath.

## See also

> ▸ Refer to the *Using Hadoop with legacy applications - Hadoop streaming* and *Adding support for new input data formats - implementing a custom InputFormat* recipes of *Chapter 4, Developing Complex Hadoop MapReduce Applications*.

# De-duplicating data using Hadoop streaming

Often, the datasets contain duplicate items that need to be eliminated to ensure the accuracy of the results. In this recipe, we use Hadoop to remove the duplicate mail records in the 20news dataset. These duplicate records are due to the users cross-posting the same message to multiple newsboards.

## Getting ready

> ▸ Make sure Python is installed on your Hadoop compute nodes.

## How to do it...

The following steps show how to remove duplicate mails due to cross-posting across the lists, from the 20news dataset:

1. Download and extract the 20news dataset from `http://qwone.com/~jason/20N ewsgroups/20news-19997.tar.gz`:

   ```
   $ wget http://qwone.com/~jason/20Newsgroups/20news-19997.tar.gz
   ```

   ```
   $ tar -xzf 20news-19997.tar.gz
   ```

2. Upload the extracted data to the HDFS. In order to save the compute time and resources, you can use only a subset of the dataset:

   ```
   $ hdfs dfs -mkdir 20news-all
   ```

   ```
   $ hdfs dfs -put  <extracted_folder> 20news-all
   ```

3. We are going to use the `MailPreProcessor.py` Python script from the previous recipe, *Data preprocessing using Hadoop streaming and Python* as the Mapper. Locate the `MailPreProcessorReduce.py` file in the source repository of this chapter.

4. Execute the following command:

```
$ hadoop jar \
/usr/lib/hadoop-mapreduce/hadoop-streaming.jar \
-input 20news-all/*/* \
-output 20news-dedup\
-mapper MailPreProcessor.py \
-reducer MailPreProcessorReduce.py \
-file MailPreProcessor.py\
-file MailPreProcessorReduce.py
```

5. Inspect the results using the following command:

```
$ hdfs dfs -cat 20news-dedup/part-00000 | more
```

## How it works...

The Mapper Python script outputs the MessageID as the key. We use the MessageID to identify the duplicated messages that are a result of cross-posting across different newsgroups.

Hadoop streaming provides the Reducer input records of each key group line by line to the streaming reducer application through the standard input. However, Hadoop streaming does not have a mechanism to distinguish a new key-value group. The streaming reducer applications need to keep track of the input key to identify new groups when Hadoop starts to feed records of a new Key to the process. Since we output the Mapper results using the MessageID, the Reducer input gets grouped by the MessageID. Any group with more than one value (aka a message) per MessageID contains duplicates. In the following script, we use only the first value (message) of the record group and discard the others, which are the duplicate messages:

```python
#!/usr/bin/env python
import sys;

currentKey = ""

for line in sys.stdin:
  line = line.strip()
  key, value = line.split('\t',1)
  if currentKey == key :
    continue
  print '%s\t%s' % (key, value)
```

See also

▸ The *Using Hadoop with legacy applications – Hadoop streaming* recipe of *Chapter 4, Developing Complex Hadoop MapReduce Applications* and the *Data preprocessing using Hadoop streaming and Python* recipe of this chapter.

# Loading large datasets to an Apache HBase data store – importtsv and bulkload

The Apache HBase data store is very useful when storing large-scale data in a semi-structured manner, so that it can be used for further processing using Hadoop MapReduce programs or to provide a random access data storage for client applications. In this recipe, we are going to import a large text dataset to HBase using the `importtsv` and `bulkload` tools.

## Getting ready

1.  Install and deploy Apache HBase in your Hadoop cluster.
2.  Make sure Python is installed in your Hadoop compute nodes.

## How to do it...

The following steps show you how to load the TSV (tab-separated value) converted 20news dataset in to an HBase table:

1.  Follow the *Data preprocessing using Hadoop streaming and Python* recipe to perform the preprocessing of data for this recipe. We assume that the output of the following step 4 of that recipe is stored in an HDFS folder named "`20news-cleaned`":

    ```
    $ hadoop jar \
        /usr/lib/hadoop-mapreduce/hadoop-streaming.jar \
        -input 20news-all/*/* \
        -output 20news-cleaned \
        -mapper MailPreProcessor.py \
        -file MailPreProcessor.py
    ```

2.  Start the HBase shell:

    ```
    $ hbase shell
    ```

3. Create a table named 20news-data by executing the following command in the HBase shell. Older versions of the `importtsv` (used in the next step) command can handle only a single column family. Hence, we are using only a single column family when creating the HBase table:

```
hbase(main):001:0> create '20news-data','h'
```

4. Execute the following command to import the preprocessed data to the HBase table created earlier:

```
$ hbase \
   org.apache.hadoop.hbase.mapreduce.ImportTsv \
   -Dimporttsv.columns=HBASE_ROW_KEY,h:from,h:group,h:subj,h:msg \
   20news-data 20news-cleaned
```

5. Start the HBase Shell and use the count and scan commands of the HBase shell to verify the contents of the table:

```
hbase(main):010:0> count '20news-data'

 12xxx row(s) in 0.0250 seconds


hbase(main):010:0> scan '20news-data', {LIMIT => 10}
 ROW                                    COLUMN+CELL
 <1993Apr29.103624.1383@cronkite.ocis.te column=h:c1,
  timestamp=1354028803355, value= katop@astro.ocis.temple.edu
  (Chris Katopis)>
 <1993Apr29.103624.1383@cronkite.ocis.te column=h:c2,
  timestamp=1354028803355, value= sci.electronics
 . . . . . .
```

The following are the steps to load the `20news` dataset to an HBase table using the `bulkload` feature:

1. Follow steps 1 to 3, but create the table with a different name:

```
hbase(main):001:0> create '20news-bulk','h'
```

2. Use the following command to generate an HBase `bulkload` datafile:

```
$ hbase \
   org.apache.hadoop.hbase.mapreduce.ImportTsv \
   -Dimporttsv.columns=HBASE_ROW_KEY,h:from,h:group,h:subj,h:msg\
   -Dimporttsv.bulk.output=hbaseloaddir \
   20news-bulk-source 20news-cleaned
```

3. List the files to verify that the `bulkload` datafiles are generated:

```
$ hadoop fs -ls 20news-bulk-source
......
drwxr-xr-x   - thilina supergroup          0 2014-04-27 10:06 /
user/thilina/20news-bulk-source/h

$ hadoop fs -ls 20news-bulk-source/h
-rw-r--r--   1 thilina supergroup      19110 2014-04-27 10:06 /
user/thilina/20news-bulk-source/h/4796511868534757870
```

4. The following command loads the data to the HBase table by moving the output files to the correct location:

```
$ hbase \
  org.apache.hadoop.hbase.mapreduce.LoadIncrementalHFiles \
  20news-bulk-source 20news-bulk
......
14/04/27 10:10:00 INFO mapreduce.LoadIncrementalHFiles: Trying
to load hfile=hdfs://127.0.0.1:9000/user/thilina/20news-bulk-
source/h/4796511868534757870 first= <1993Apr29.103624.1383@
cronkite.ocis.temple.edu>last= <stephens.736002130@ngis>
......
```

5. Start the HBase Shell and use the `count` and `scan` commands of the HBase shell to verify the contents of the table:

```
hbase(main):010:0> count '20news-bulk'
hbase(main):010:0> scan '20news-bulk', {LIMIT => 10}
```

## How it works...

The `MailPreProcessor.py` Python script extracts a selected set of data fields from the newsboard message and outputs them as a tab-separated dataset:

```
value = fromAddress + "\t" + newsgroup
+"\t" + subject +"\t" + value
print '%s\t%s' % (messageID, value)
```

We import the tab-separated dataset generated by the Streaming MapReduce computations to HBase using the `importtsv` tool. The `importtsv` tool requires the data to have no other tab characters except for the tab characters that separate the data fields. Hence, we remove any tab characters that may be present in the input data by using the following snippet of the Python script:

```
line = line.strip()
line = re.sub('\t',' ',line)
```

The `importtsv` tool supports the loading of data into HBase directly using the `Put` operations as well as by generating the HBase internal `HFiles` as well. The following command loads the data to HBase directly using the `Put` operations. Our generated dataset contains a Key and four fields in the values. We specify the data fields to the table column name mapping for the dataset using the `-Dimporttsv.columns` parameter. This mapping consists of listing the respective table column names in the order of the tab-separated data fields in the input dataset:

```
$ hbase \
   org.apache.hadoop.hbase.mapreduce.ImportTsv \
   -Dimporttsv.columns=<data field to table column mappings> \
   <HBase tablename> <HDFS input directory>
```

We can use the following command to generate HBase HFiles for the dataset. These HFiles can be directly loaded to HBase without going through the HBase APIs, thereby reducing the amount of CPU and network resources needed:

```
$ hbase \
   org.apache.hadoop.hbase.mapreduce.ImportTsv \
   -Dimporttsv.columns=<filed to column mappings> \

   -Dimporttsv.bulk.output=<path for hfile output> \
   <HBase tablename> <HDFS input directory>
```

These generated HFiles can be loaded into HBase tables by simply moving the files to the right location. This moving can be performed by using the `completebulkload` command:

```
$ hbase \
   org.apache.hadoop.hbase.mapreduce.LoadIncrementalHFiles \

   <HDFS path for hfiles> <table name>
```

## There's more...

You can use the `importtsv` tool that has datasets with other data-filed separator characters as well by specifying the '-Dimporttsv.separator' parameter. The following is an example of using a comma as the separator character to import a comma-separated dataset in to an HBase table:

```
$ hbase \
   org.apache.hadoop.hbase.mapreduce.ImportTsv \
   '-Dimporttsv.separator=,' \

   -Dimporttsv.columns=<data field to table column mappings> \
   <HBase tablename> <HDFS input directory>
```

Look out for `Bad Lines` in the MapReduce job console output or in the Hadoop monitoring console. One reason for `Bad Lines` is to have unwanted delimiter characters. The Python script we used in the data-cleaning step removes any extra tabs in the message:

```
14/03/27 00:38:10 INFO mapred.JobClient:    ImportTsv
14/03/27 00:38:10 INFO mapred.JobClient:      Bad Lines=2
```

## Data de-duplication using HBase

HBase supports the storing of multiple versions of column values for each record. When querying, HBase returns the latest version of values, unless we specifically mention a time period. This feature of HBase can be used to perform automatic de-duplication by making sure we use the same RowKey for duplicate values. In our 20news example, we use MessageID as the RowKey for the records, ensuring duplicate messages will appear as different versions of the same data record.

HBase allows us to configure the maximum or minimum number of versions per column family. Setting the maximum number of versions to a low value will reduce the data usage by discarding the old versions. Refer to `http://hbase.apache.org/book/schema.versions.html` for more information on setting the maximum or minimum number of versions.

## See also

▸ The *Running MapReduce jobs on HBase* recipe of *Chapter 7, Hadoop Ecosystem II – Pig, HBase, Mahout, and Sqoop*.

▸ Refer to `http://hbase.apache.org/book/ops_mgt.html#importtsv` for more information on the `ImportTsv` command.

# Creating TF and TF-IDF vectors for the text data

Most of the text analysis data-mining algorithms operate on vector data. We can use a vector space model to represent text data as a set of vectors. For example, we can build a vector space model by taking the set of all terms that appear in the dataset and by assigning an index to each term in the term set. The number of terms in the term set is the dimensionality of the resulting vectors, and each dimension of the vector corresponds to a term. For each document, the vector contains the number of occurrences of each term at the index location assigned to that particular term. This creates the vector space model using term frequencies in each document, which is similar to the result of the computation we performed in the *Generating an inverted index using Hadoop MapReduce* recipe of *Chapter 8, Searching and Indexing*.

The vectors can be seen as follows:

The term frequencies and the resulting document vectors

However, creating vectors using the preceding term count model gives a lot of weight to the terms that occur frequently across many documents (for example, the, is, a, are, was, who, and so on), although these frequent terms have a very minimal contribution when it comes to defining the meaning of a document. The **Term frequency-inverse document frequency (TF-IDF)** model solves this issue by utilizing the **inverted document frequencies (IDF)** to scale the **term frequencies (TF)**. IDF is typically calculated by first counting the number of documents (DF) the term appears in, inversing it (1/DF) and normalizing it by multiplying with the number of documents and using the logarithm of the resultant value as shown roughly by the following equation:

$$ \text{TF-IDF}_i = \text{TF}_i \times \log (N/DF_i) $$

In this recipe, we'll create TF-IDF vectors from a text dataset using a built-in utility tool of Apache Mahout.

## Getting ready

Install Apache Mahout in your machine using your Hadoop distribution or install the latest Apache Mahout version manually.

## How to do it...

The following steps show you how to build a vector model of the 20news dataset:

1.  Download and extract the 20news dataset from `http://qwone.com/~jason/20N ewsgroups/20news-19997.tar.gz`:

    ```
    $ wget http://qwone.com/~jason/20Newsgroups/20news-19997.tar.gz
    $ tar -xzf 20news-19997.tar.gz
    ```

2.  Upload the extracted data to the HDFS. In order to save the compute time and resources, you may use only a subset of the dataset:

    ```
    $ hdfs dfs -mkdir 20news-all
    $ hdfs dfs -put  <extracted_folder> 20news-all
    ```

3.  Go to MAHOUT_HOME. Generate the Hadoop sequence files from the uploaded text data:

    ```
    $ mahout seqdirectory -i 20news-all -o 20news-seq
    ```

4.  Generate TF and TF-IDF sparse vector models from the text data in the sequence files:

    ```
    $ mahout seq2sparse -i 20news-seq  -o 20news-vector
    ```

    The preceding command launches a series of MapReduce computations. Wait for the completion of these computations:

| | | | | |
|---|---|---|---|---|
| application_1398781826629_0011 | tgunarathne | PartialVectorMerger::MergePartialVectors | default | Tue, 13 May 2014 01:27:09 GMT |
| application_1398781826629_0010 | tgunarathne | : MakePartialVectors: input-folder: 20news-vector/tf-vectors, dictionary-file: 20news-vector/frequency.file-0 | default | Tue, 13 May 2014 01:26:05 GMT |
| application_1398781826629_0009 | tgunarathne | VectorTfIdf Document Frequency Count running over input: 20news-vector/tf-vectors | default | Tue, 13 May 2014 01:25:05 GMT |
| application_1398781826629_0008 | tgunarathne | PartialVectorMerger::MergePartialVectors | default | Tue, 13 May 2014 01:23:59 GMT |
| application_1398781826629_0007 | tgunarathne | DictionaryVectorizer::MakePartialVectors: input-folder: 20news-vector/tokenized-documents, dictionary-file: 20news-vector/dictionary.file-0 | default | Tue, 13 May 2014 01:22:30 GMT |
| application_1398781826629_0006 | tgunarathne | DictionaryVectorizer::WordCount: input-folder: 20news-vector/tokenized-documents | default | Tue, 13 May 2014 01:20:48 GMT |
| application_1398781826629_0005 | tgunarathne | DocumentProcessor::DocumentTokenizer: input-folder: 20news-seq | default | Tue, 13 May 2014 |

5. Check the output directory using the following command. The `tfidf-vectors` folder contains the TF-IDF model vectors, the `tf-vectors` folder contains the term count model vectors, and the `dictionary.file-0` contains the term to term-index mapping:

```
$ hdfs dfs -ls 20news-vector
```

```
Found 7 items
drwxr-xr-x   - u supergroup          0 2012-11-27 16:53 /user/
u/20news-vector /df-count
-rw-r--r--   1 u supergroup       7627 2012-11-27 16:51 /user/
u/20news-vector/dictionary.file-0
-rw-r--r--   1 u supergroup       8053 2012-11-27 16:53 /user/
u/20news-vector/frequency.file-0
drwxr-xr-x   - u supergroup          0 2012-11-27 16:52 /user/
u/20news-vector/tf-vectors
drwxr-xr-x   - u supergroup          0 2012-11-27 16:54 /user/
u/20news-vector/tfidf-vectors
drwxr-xr-x   - u supergroup          0 2012-11-27 16:50 /user/
u/20news-vector/tokenized-documents
drwxr-xr-x   - u supergroup          0 2012-11-27 16:51 /user/
u/20news-vector/wordcount
```

6. Optionally, you can use the following command to dump the TF-IDF vectors as text. The key is the filename and the contents of the vectors are in the format `<term index>:<TF-IDF value>`:

```
$ mahout seqdumper -i 20news-vector/tfidf-vectors/part-r-00000
```

```
......

Key class: class org.apache.hadoop.io.Text Value Class: class org.
apache.mahout.math.VectorWritable
Key: /54492: Value: {225:3.374729871749878,400:1.5389964580535889,
321:1.0,324:2.386294364929199,326:2.386294364929199,315:1.0,144:2.
0986123085021973,11:1.0870113372802734,187:2.652313232421875,134:2
.386294364929199,132:2.0986123085021973,......}

......
```

## How it works...

Hadoop SequenceFiles store the data as binary key-value pairs and support data compression. Mahout's `seqdirectory` command converts the text files into a Hadoop SequenceFile by using the filename of the text file as the key and the contents of the text file as the value. The `seqdirectory` command stores all the text contents in a single SequenceFile. However, it's possible for us to specify a chunk size to control the actual storage of the SequenceFile data blocks in the HDFS. The following are a selected set of options for the `seqdirectory` command:

```
mahout seqdirectory
 -i <HDFS path to text files>
 -o <HDFS output directory for sequence file>
 -ow                    If present, overwrite the output directory
 -chunk <chunk size>    In MegaBytes. Defaults to 64mb
 -prefix <key prefix>   The prefix to be prepended to the key
```

The `seq2sparse` command is an Apache Mahout tool that supports the generation of sparse vectors from SequenceFiles that contain text data. It supports the generation of both TF as well as TF-IDF vector models. This command executes as a series of MapReduce computations. The following are a selected set of options for the `seq2sparse` command:

```
mahout seq2sparse
 -i <HDFS path to the text sequence file>
 -o <HDFS output directory>
 -wt {tf|tfidf}
 -chunk <max dictionary chunk size in mb to keep in memory>
 --minSupport <minimum support>
 --minDF <minimum document frequency>
 --maxDFPercent <MAX PERCENTAGE OF DOCS FOR DF
```

The `minSupport` command is the minimum frequency for the word to be considered as a feature. `minDF` is the minimum number of documents the word needs to be in. `maxDFPercent` is the maximum value of the expression (document frequency of a word/total number of document) in order for that word to be considered as a good feature in the document. This helps remove high-frequency features such as stop words.

You can use the Mahout `seqdumper` command to dump the contents of a SequenceFile that uses the Mahout writable data types as plain text:

```
mahout seqdumper
 -i <HDFS path to the sequence file>
 -o <output directory>
 --count        Output only the number of key value pairs.
 --numItems     Max number of key value pairs to output
 --facets       Output the counts per key.
```

## See also

▶ The *Generating an inverted index using Hadoop MapReduce* recipe of *Chapter 9, Classifications, Recommendations, and Finding Relationships.*

▶ Refer to the Mahout documentation on creating vectors from text data at `https://cwiki.apache.org/confluence/display/MAHOUT/Creating+Vectors+from+Text`.

# Clustering text data using Apache Mahout

Clustering plays an integral role in data-mining computations. Clustering groups together similar items of a dataset using one or more features of the data items based on the use case. Document clustering is used in many text-mining operations such as document organization, topic identification, information presentation, and so on. Document clustering shares many of the mechanisms and algorithms with traditional data clustering mechanisms. However, document clustering has its unique challenges when it comes to determining the features to use for clustering and when building vector space models to represent the text documents.

The *Running K-means with Mahout* recipe of *Chapter 7, Hadoop Ecosystem II – Pig, HBase, Mahout, and Sqoop* focuses on using Mahout KMeansClustering to cluster a statistics data. The *Clustering an Amazon sales dataset* recipe of *Chapter 8, Classifications, Recommendations, and Finding Relationships* of the previous edition of this book focuses on using clustering to identify customers with similar interests. These two recipes provide a more in-depth understanding of using Clustering algorithms in general. This recipe focuses on exploring two of the several clustering algorithms available in Apache Mahout for document clustering.

## Getting ready

▶ Install Apache Mahout in your machine using your Hadoop distribution or install the latest Apache Mahout version manually in your machine.

## How to do it...

The following steps use the Apache Mahout KmeansClustering algorithm to cluster the 20news dataset:

1. Refer to the *Creating TF and TF-IDF vectors for the text data* recipe in this chapter and generate TF-IDF vectors for the 20news dataset. We assume the TF-IDF vectors are in the `20news-vector/tfidf-vectors` folder of HDFS.

2.  Execute the following command to run the Mahout KMeansClustering computation:

```
$ mahout kmeans \
  --input 20news-vector/tfidf-vectors \
  --clusters 20news-seed/clusters
  --output 20news-km-clusters\
  --distanceMeasure \
org.apache.mahout.common.distance.SquaredEuclideanDistanceMeasure
  -k 10 --maxIter 20 --clustering
Execute the following command to convert the clusters to text:
$ mahout clusterdump \
  -i 20news-km-clusters/clusters-*-final\
  -o 20news-clusters-dump \
  -d 20news-vector/dictionary.file-0 \
  -dt sequencefile \
  --pointsDir 20news-km-clusters/clusteredPoints


$ cat 20news-clusters-dump
```

## How it works...

The following code shows the usage of the Mahout KMeans algorithm:

```
mahout kmeans
  --input <tfidf vector input>
  --clusters <seed clusters>
  --output <HDFS path for output>
  --distanceMeasure <distance measure>
  -k <number of clusters>
  --maxIter <maximum number of iterations>
  --clustering
```

Mahout will generate random seed clusters when an empty HDFS directory path is given to the --clusters option. Mahout supports several different distance calculation methods such as Euclidean, Cosine, and Manhattan.

The following is the usage of the Mahout `clusterdump` command:

```
mahout clusterdump
   -i <HDFS path to clusters>
   -o <local path for text output>
   -d <dictionary mapping for the vector data points>
   -dt <dictionary file type (sequencefile or text)>
   --pointsDir <directory containing the input vectors to
                  clusters mapping>
```

## See also

   ▶  The *Running K-means with Mahout* recipe of *Chapter 7, Hadoop Ecosystem II – Pig, HBase, Mahout, and Sqoop*.

# Topic discovery using Latent Dirichlet Allocation (LDA)

We can use **Latent Dirichlet Allocation** (**LDA**) to cluster a given set of words into topics and a set of documents into combinations of topics. LDA is useful when identifying the meaning of a document or a word based on the context, without solely depending on the number of words or the exact words. LDA is a step away from raw text matching and towards semantic analysis. LDA can be used to identify the intent and to resolve ambiguous words in a system such as a search engine. Some other example use cases of LDA are identifying influential Twitter users for particular topics and Twahpic (`http://twahpic.cloudapp.net`) application uses LDA to identify topics used on Twitter.

LDA uses the TF vector space model as opposed to the TF-IDF model as it needs to consider the co-occurrence and correlation of words.

## Getting ready

Install Apache Mahout in your machine using your Hadoop distribution, or install the latest Apache Mahout version manually.

## How to do it...

The following steps show you how to run the Mahout LDA algorithm on a subset of the 20news dataset:

1. Download and extract the 20news dataset from `http://qwone.com/~jason/20N ewsgroups/20news-19997.tar.gz`:

   ```
   $ wget http://qwone.com/~jason/20Newsgroups/20news-19997.tar.gz
   $ tar -xzf 20news-19997.tar.gz
   ```

2. Upload the extracted data to the HDFS. In order to save the compute time and resources, you may use only a subset of the dataset:

   ```
   $ hdfs dfs -mkdir 20news-all
   $ hdfs dfs -put  <extracted_folder> 20news-all
   ```

3. Generate sequence files from the uploaded text data:

   ```
   $ mahout seqdirectory -i 20news-all -o 20news-seq
   ```

4. Generate a sparse vector from the text data in the sequence files:

   ```
   $ mahout seq2sparse \
   -i 20news-seq  -o 20news-tf \
   -wt tf -a org.apache.lucene.analysis.WhitespaceAnalyzer
   ```

5. Convert the TF vectors from SequenceFile<Text, VectorWritable> to SequenceFile<IntWritable,Text>:

   ```
   $ mahout rowid -i 20news-tf/tf-vectors -o 20news-tf-int
   ```

6. Run the following command to perform the LDA computation:

   ```
   $ mahout cvb \
   -i 20news-tf-int/matrix -o lda-out \
   -k 10  -x 20  \
   -dict 20news-tf/dictionary.file-0 \
   -dt lda-topics \
   -mt lda-topic-model
   ```

7. Dump and inspect the results of the LDA computation:

   ```
   $ mahout seqdumper -i lda-topics/part-m-00000


   Input Path: lda-topics5/part-m-00000
   ```

```
Key class: class org.apache.hadoop.io.IntWritable Value Class:
class org.apache.mahout.math.VectorWritable

Key: 0: Value: {0:0.12492744375758073,1:0.03875953927132082,2:0.12
28639250669511,3:0.15074522974495433,4:0.10512715697420276,5:0.101
30565323653766,6:0.061169131590630275,7:0.14501579630233746,8:0.07
872957132697946,9:0.07135655272850545}
```

. . . . .

8. Join the output vectors with the dictionary mapping of term-to-term indexes:

```
$ mahout vectordump \
-i lda-topics/part-m-00000 \
--dictionary 20news-tf/dictionary.file-0 \
--vectorSize 10   -dt sequencefile
```

. . . . . .

```
{"Fluxgate:0.12492744375758073,&:0.03875953927132082,(140.220.1.1
):0.1228639250669511,(Babak:0.15074522974495433,(Bill:0.105127156
97420276,(Gerrit:0.10130565323653766,(Michael:0.06116913159063027
5,(Scott:0.14501579630233746,(Usenet:0.07872957132697946,(continu
ed):0.07135655272850545}

{"Fluxgate:0.13130952097888746,&:0.05207587369196414,(140.220.1.1
):0.12533225607394424,(Babak:0.08607740024552457,(Bill:0.20218284
543514245,(Gerrit:0.07318295757631627,(Michael:0.0876688824220103
9,(Scott:0.08858421220476514,(Usenet:0.09201906604666685,(continu
ed):0.06156698532477829}
```

. . . . . . .

## How it works...

The Mahout CVB version of LDA implements the Collapse Variable Bayesian inference algorithm using an iterative MapReduce approach:

```
mahout cvb \
-i 20news-tf-int/matrix \
-o lda-out -k 10   -x 20 \
-dict 20news-tf/dictionary.file-0 \
-dt lda-topics \
-mt lda-topic-model
```

The -i parameter provides the input path, while the -o parameter provides the path to store the output. The -k parameter specifies the number of topics to learn and −x specifies the maximum number of iterations for the computation. The -dict parameter points to the dictionary that contains the mapping of terms to term-indexes. The path given in the −dt parameter stores the training topic distribution. The path given in −mt is used as a temporary location to store the intermediate models.

All the command-line options of the cvb command can be queried by invoking the help option as follows:

```
mahout  cvb  --help
```

Setting the number of topics to a very small value brings out extremely high-level topics. A large number of topics produces more descriptive topics but takes longer to process. The maxDFPercent option can be used to remove common words, thereby speeding up the processing.

## See also

► *A Collapsed Variational Bayesian Inference Algorithm for Latent Dirichlet Allocation* by Y.W. Teh, D. Newman, and M. Welling. In NIPS, volume 19, 2006 which can be found at http://www.gatsby.ucl.ac.uk/~ywteh/research/inference/nips2006.pdf.

# Document classification using Mahout Naive Bayes Classifier

Classification assigns documents or data items to an already known set of classes with already known properties. Document classification or categorization is used when we need to assign documents to one or more categories. This is a frequent use case in information retrieval as well as library science.

The *Classification using the naïve Bayes classifier* recipe in *Chapter 9, Classifications, Recommendations, and Finding Relationships* provides a more detailed description about classification use cases, and also gives you an overview of using the Naive Bayes classifier algorithm. This recipe focuses on highlighting the classification support in Apache Mahout for text documents.

## Getting ready

► Install Apache Mahout in your machine using your Hadoop distribution, or install the latest Apache Mahout version manually.

## How to do it...

The following steps use the Apache Mahout Naive Bayes algorithm to cluster the 20news dataset:

1. Refer to the *Creating TF and TF-IDF vectors for the text data* recipe in this chapter and generate TF-IDF vectors for the 20news dataset. We assume that the TF-IDF vectors are in the `20news-vector/tfidf-vectors` folder of the HDFS.

2. Split the data into training and test datasets:

```
$ mahout split \
    -i 20news-vectors/tfidf-vectors \
    --trainingOutput /20news-train-vectors \
    --testOutput /20news-test-vectors  \
    --randomSelectionPct 40 \
--overwrite --sequenceFiles
```

3. Train the Naive Bayes model:

```
$ mahout trainnb \
    -i 20news-train-vectors -el \
    -o  model \
    -li labelindex
```

4. Test the classification on the test dataset:

```
$ mahout testnb \
     -i 20news-train-vectors \
    -m model \
    -l labelindex \
    -o 20news-testing
```

## How it works...

Mahout's `split` command can be used to split a dataset into a training dataset and a test dataset. This command works with text datasets as well as with Hadoop SequenceFile datasets. The following is the usage of the Mahout `data-splitting` command. You can use the `--help` option with the `split` command to print out all the options:

```
mahout split \
  -i <input data directory> \
```

```
--trainingOutput <HDFS path to store the training dataset> \
--testOutput <HDFS path to store the test dataset>  \
--randomSelectionPct <percentage to be selected as test data> \
--sequenceFiles
```

The `sequenceFiles` option specifies that the input dataset is in Hadoop SequenceFiles.

The following is the usage of the Mahout Naive Bayes classifier training command. The `--el` option informs Mahout to extract the labels from the input dataset:

```
mahout trainnb \
  -i <HDFS path to the training data set> \
  -el \
  -o <HDFS path to store the trained classifier model> \
  -li <Path to store the label index> \
```

The following is the usage of the Mahout Naive Bayes classifier testing command:

```
mahout testnb \
    -i <HDFS path to the test data set>
    -m <HDFS path to the classifier model>\
    -l <Path to the label index> \
    -o <path to store the test result>
```

## See also

▶ The *Classification using the naïve Bayes classifier* recipe of *Chapter 9, Classifications, Recommendations, and Finding Relationships*

# Index

# W

# Y

## Thank you for buying
## Hadoop MapReduce v2 Cookbook
### Second Edition

# About Packt Publishing

Packt, pronounced 'packed', published its first book, *Mastering phpMyAdmin for Effective MySQL Management*, in April 2004, and subsequently continued to specialize in publishing highly focused books on specific technologies and solutions.

Our books and publications share the experiences of your fellow IT professionals in adapting and customizing today's systems, applications, and frameworks. Our solution-based books give you the knowledge and power to customize the software and technologies you're using to get the job done. Packt books are more specific and less general than the IT books you have seen in the past. Our unique business model allows us to bring you more focused information, giving you more of what you need to know, and less of what you don't.

Packt is a modern yet unique publishing company that focuses on producing quality, cutting-edge books for communities of developers, administrators, and newbies alike. For more information, please visit our website at www.packtpub.com.

# About Packt Open Source

In 2010, Packt launched two new brands, Packt Open Source and Packt Enterprise, in order to continue its focus on specialization. This book is part of the Packt open source brand, home to books published on software built around open source licenses, and offering information to anybody from advanced developers to budding web designers. The Open Source brand also runs Packt's open source Royalty Scheme, by which Packt gives a royalty to each open source project about whose software a book is sold.

# Writing for Packt

We welcome all inquiries from people who are interested in authoring. Book proposals should be sent to author@packtpub.com. If your book idea is still at an early stage and you would like to discuss it first before writing a formal book proposal, then please contact us; one of our commissioning editors will get in touch with you.

We're not just looking for published authors; if you have strong technical skills but no writing experience, our experienced editors can help you develop a writing career, or simply get some additional reward for your expertise.

## Optimizing Hadoop for MapReduce

ISBN: 978-1-78328-565-5          Paperback: 120 pages

Learn how to configure your Hadoop cluster to run optimal MapReduce jobs

1. Optimize your MapReduce job performance.

2. Identify your Hadoop cluster's weaknesses.

3. Tune your MapReduce configuration.

## Hadoop Real-World Solutions Cookbook

ISBN: 978-1-84951-912-0          Paperback: 316 pages

Realistic, simple code examples to solve problems at scale with Hadoop and related technologies

1. Solutions to common problems when working in the Hadoop environment.

2. Recipes for (un)loading data, analytics, and troubleshooting.

3. In-depth code examples demonstrating various analytic models, analytic solutions, and common best practices.

Please check **www.PacktPub.com** for information on our titles

## Big Data Analytics with R and Hadoop

ISBN: 978-1-78216-328-2          Paperback: 238 pages

Set up an integrated infrastructure of R and Hadoop to turn your data analytics into Big Data analytics

1. Write Hadoop MapReduce within R.

2. Learn data analytics with R and the Hadoop platform.

3. Handle HDFS data within R.

4. Understand Hadoop streaming with R.

## Instant MapReduce Patterns – Hadoop Essentials How-to

ISBN: 978-1-78216-770-9          Paperback: 60 pages

Practical recipes to write your own MapReduce solution patterns for Hadoop programs

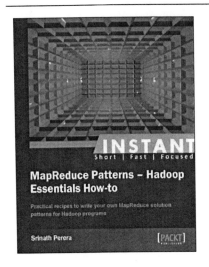

1. Learn something new in an Instant! A short, fast, focused guide delivering immediate results.

2. Learn how to install, configure, and run Hadoop jobs.

3. Seven recipes, each describing a particular style of the MapReduce program to give you a good understanding of how to program with MapReduce.

Please check **www.PacktPub.com** for information on our titles

CPSIA information can be obtained
at www.ICGtesting.com
Printed in the USA
FFOW03n0054020315
11452FF